GVL怡境景观

Greenview Landscape

GVL 国际怡境景观设计有限公司、佳图文化　编

www.gvlhk.com

广州总部
地址：珠江新城华夏路 49 号津滨腾越大厦南塔 8-9 楼
邮编：510623
电话：(020)38032695、(020)38032762
传真：(020)38032716
邮箱：GVL@gvlcn.com

Guangzhou
8-9/F, South Tower, Jinbin Tengyue Building, No.49, Hua Xia Road, Zhujiang New Town, Guangzhou, 510623, P.R.China
Tel: (020)38032695、(020)38032762
Fax: (020)38032716
E-Mail: GVL@gvlcn.com

香港
地址：北角渣华道 18 号嘉汇商业大厦 2106 室
传真：(852)22934388
邮箱编号：070432

Hong Kong
Room 2106, Carnival Commercial Building, 18 Java Road, North Point, Hong Kong
Fax: (852)22934388
Mailbox: 070432

深圳
地址：福田区深南中路田面设计之都创意产业园 5 栋 2 层 A1
邮编：518026
电话：(0755)23931053
传真：(0755)23931053-818
邮箱：gvl-sz@greenview.com.cn

Shenzhen
Room A1, 2nd Floor, Building 5, Tianmian City of Design Industrial Estate, Shen Nan Zhong Road, Futian District, Shenzhen, 518026, P.R.China
Tel: (0755)23931053
Fax: (0755)23931053-818
E-Mail: gvl-sz@greenview.com.cn

长沙
地址：雨花区湘府东路二段 108 号水岸天际第 1 栋写字楼 1619-1625 室
邮编：410004
电话：(0731)89855708、（0731）89855718
传真：(0731)89855728
邮箱：gvlcncs@163.com

Changsha
Room 1619-1625, No 1 Office Building, Park Linking Community, No.108, Xiangfu East Road Section Ⅱ, Yuhua District, Changsha, 410004, P.R.China
Tel: (0731)89855708、(0731)89855718
Fax: (0731)89855728
E-Mail: gvlcncs@163.com

北京
地址：朝阳区望京ＳＯＨＯ１号塔楼Ｂ座 1907 室
邮编：100101
电话：(010)57076981
传真：(010)57076981-818
邮箱：gvlcnbj@163.com

Beijing
Room 1907, Block B, Tower 1, Wangjing SOHO, Chaoyang District, Beijing, 100101, P.R.China
Tel: (010)57076981
Fax: (010)57076981-818
E-Mail: gvlcnbj@163.com

上海
地址：长宁区愚园路 1107 号 1 号楼 4 楼创邑·米域
邮编：200050

Shanghai
MIXpace,4th Floor, Block 1, NO.1107, Yuyuan Road, Changning District, Shanghai, 200050,P.R.China

GVL怡境景观

Greenview Landscape

GVL 国际怡境景观设计有限公司、佳图文化　编

Dedication and Determination lead to Success

With wisdom and efforts of all GVL people, the Collection of GVL Works is finally published. As a special "gift" for the 15th anniversary of GVL, it has collected GVL's classic works in recent years and shown GVL people's dreams of the future.

Since its foundation, GVL has created about 1000 projects in more than 100 cities, which carry our memories and have their own characteristics. Therefore, it is not easy for us to make a selection from them.

Though 15 years is just a moment in the long history of time, it is a critical period for further development of an enterprise. After 15 years of steady development, GVL has now become a professional and world-renowned landscape design company, whose works have won BALI (British Association of Landscape Industries) International Gold Award twice as well as other prizes and honors. What makes us even more excited is that GVL's works are widely received by the market.

People often asked me about the secret of GVL's growth and success. Simply speaking, it is due to three elements: talents, team and culture. Talents form a team, and then shape the team culture. Peter Ferdinand Drucker, the inventor of modern management has said, for any organization, " the key to greatness is to look for people's potential and spend time developing it."

In fact, every one has his own expectation for the future. During our career, we will always think about "who we want to be" and "what our company will grow like." As the leader of the GVL team, I always insist on two things: good management and good design. With clear goal and firm determination, I will devote myself to building the platform where GVL people can realize their dreams.

Beijing Tongrentang, a famous time-honored brand in traditional Chinese medicine industry, has been adhering to the full significance of the handed down TRT's commandment: "No manpower shall be spared, no matter how complicated the procedures of pharmaceutical production; and no material shall be reduced, no matter how much the cost." Throughout history, we will find that almost all excellent enterprises have one thing in common: they are focusing on what they are doing and ensuring the quality of their products or service. In GVL, we always do the same: we devote ourselves to presenting high-quality works and we treasure every design opportunity, no matter whether the project is big or small, profitable or not.

Dedication and Determination lead to Success. GVL keeps the faith to design well and to present classic works, and we have the confidence to be outstanding in the landscape design industry. After 15 years of development, now GVL is heading into a new era. With beautiful dreams in mind, we will walk forward step by step!

By Peter Peng , Managing Director, Senior Engineer, Greenview Landscape International Design Group (GVL)

保利

勿忘初心，方得始终

GVL 怡境景观的《作品集》终于要面世了，这本汇聚了怡境人智慧结晶，收录了怡境近年经典项目的作品集，将作为怡境公司十五周年的献礼，承载着怡境人对未来的憧憬展现在世人面前。

GVL 怡境景观成立至今，作品遍布过百个城市，项目近千，要从中选出若干作品集结成册，着实不易，因为每个作品都有其专属回忆，都有其闪光点，让人难以取舍。

在历史的长河中，十五载不过是弹指一瞬，而在一个企业的成长史上，十五年正是迈向崭新时代的关键时期。十五年来，GVL 怡境景观稳健发展，成长为国内外知名的专业景观设计公司，两度荣膺英国国家园景工业协会（BALI）国际园景项目金奖，亦数次获得其他荣誉，更宝贵的是赢得了市场的广泛认同。

常有人问我 GVL 成长的秘诀是什么，简单来说，就是：人才 + 团队 + 文化。人才组成团队，团队形成文化。管理大师德鲁克说过："对于任何组织而言，伟大的关键在于寻找人的潜能，并花时间开发潜能，让凡人做非凡之事。"

事实上，每个人都对未来有着独特的期许，从业过程中，大家也会不断思索，我们究竟想成为什么样的人，我们的公司将成长为什么样的公司。作为怡境的领队，我始终坚持两个目标：管理好公司，做好设计。清晰定位、坚守本业，为怡境人提供实现梦想的平台。

"品味虽贵必不敢减物力，炮制虽烦必不敢省人工"，同仁堂凭借这条古训得以屹立百年。回顾历史，你会发现，几乎所有的优质企业，都有一个共同的特点：专注做事，坚守品质。怡境正是这样，从来都是心无旁骛地专注于精品设计，坚持信仰，沉心静气。不管设计大小，无论利润丰薄，珍惜每次创作的机会，用真材实料说话，用好作品说话。

勿忘初心，方得始终。怡境的初心，就是要做好设计，怡境追求的始终，就是要出经典作品，成为行业标杆的企业。十五年，怡境的又一个新时代来临，让我们坚持心怀梦想，仰望星空；让我们脚踏实地，迈步向前。

彭涛

GVL 怡境国际设计集团 董事长丨高级工程师

向着我们的理想迈进

Footprints on the Path of Landscape Design

There are four basic necessities of life since ancient civilisation: clothing, food, shelter and transportation. Among them, food and shelter are the primary essentials. With the improvement of living standards, more and more attention will be paid towards living conditions. Take Hong Kong as an example, in the 1950s whole families had to huddle in old dilapidated apartments under poor housing conditions. As the economy developed in the 1970s, both the Hong Kong government and private real estate developers had a surge in building public and commodity housing. At the same time, global political stability together with strong economic growth also accelerated the development of the real estate sector. Landscape architecture had emerged with this trend. At the time I just returned to Hong Kong from the United Kingdom, and I was fortunate in being able to participate in the pioneering landscape project for Swire Properties, which planned to develop a 60 acre dockyard into a modern residential town, Taikoo Shing (Taikoo City). After over 20 years of operation, the former dockyard had been transformed into one of Hong Kong's most prominent residential developments. Our Swire Properties Environment Department then evolved to be an independent environmental services company, which became the predecessor of today's GVL.

Hong Kong's Taikoo Shing is an exemplary combination of well designed landscape and residential architecture, which brought high profits for Swire Properties, hence other local developers began to follow suit. Since then, the landscape design industry in this region flourished.

In 1980s and 1990s the real estate of mainland China was booming along with the rapid economic development. Accordingly, there was increased demand for high quality landscape design. GVL seized the opportunity and shifted the focus of business to mainland China. In the following decade, GVL's business expanded swiftly with new branches established in main cities, and accomplishing projects all over China. Under sound leadership and the combined efforts of all GVL staff, by adhering to the principle of "design to establish excellent brand, service to accomplish excellent work", our projects are well received by satisfied clients. GVL is highly regarded and have won numerous national and international awards. In this collection, we have selected some exemplary projects to share with you.

Since the establishment of GVL we have been carrying out the idea of "eco-landscape design", which aims to create a high quality living environment for inhabitants. Nowadays, the declining environment and unsustainable developments have drawn the attention of many concerned governments and the public. We endeavour to preserve and improve the natural environment, reduce pollution, and save energy. Let living together in harmony with nature be our dream and goal, to advance a better environment and the people living within it.

In today's highly materialised and mechanised society, our environment is greatly polluted by noise, waste, exhaust gas and excess lighting. It seems that we are getting further and further away from nature and it is challenging to achieve sustainable development. It is crucial for landscape designers like us to always keep in mind to create well designed micro-environment which would in turn benefit our macro-environment of our whole ecosystem. This would also be part of our idealistic dream for a greater China.

By Dr WAI Tze-Kong

Chairman, Greenview Landscape International Design Group (GVL)
Chairman, Environment Committee, The General Chinese Chamber of Commerce, Hong Kong
D.Phil., University of York, U.K.
Research fellow, Department of Environment, U.K. Government

向着我们的理想迈进

从古到今，人类社会最重要的四大元素是衣食住行，而其中又以食和住为首要。随着生活水平的提高，人们对于居住条件也逐渐讲究起来。以香港为例：20 世纪 50 年代，居住条件相当落后，一家人挤在残旧楼宇中，卫生条件差，居住空间小。随着经济发展，70 年代政府大量兴建公共房屋，加以私营发展商（地产商）看到商机也大举觅地建屋。国际上也因政局稳定和经济蓬勃而促进了房地产建设，景观建筑设计应运而生。其时我自英国回港，适逢其会投身这行业，参加了太古地产公司改造太古船坞项目。该公司当时计划将 60 多英亩（24 万多平方米）的船坞所在地打造成一个现代化的具有优良人居条件的小城——太古城。经过二十多年的经营，这片“壁断垣残、一泓碧水破船湾”被改造成为当时有名的优质居所。原属太古地产的环境部亦顺应发展要求分拆成独立的环境服务公司，及后，又以此为基础在港成立怡境景观设计有限公司。

香港“太古城”树立了典范，不但给市民提供了稍具较好的景观和设计的居所，同时，发展商亦因销售迅速而大获其利。因此当地发展商便纷相效法，而景观建筑设计行业亦趋蓬勃。

至 20 世纪 80—90 年代，国内地产行业随经济发展呈兴旺势，对景观设计要求甚严，怡境亦因势利导将业务重点转移国内，十数年间，业务发展迅速，项目遍全国，分公司驻数大城市。端赖同仁努力及领导有方，本着“设计创造品牌，服务雕琢精品”宗旨，所负责的项目俱达满意水平，获业界称誉，屡得国内外奖项。其中在本刊介绍者，可见鳞爪。

怡境创建伊始便以履行生态景观设计为目标，使居民能获得全面的优质生活环境。值此地球自然环境条件日趋恶化，持续发展日趋困难，对人民生活关注的政府及人士，咸相设法维护及提升自然环境的主要元素，减少污染节省能源。返璞归真，让人们回归自然；天人合一，让人们与天地同在。

在现今高度物质化和机械化的社会，居住区的小环境往往被噪音、废物、废气甚至光电污染，人们距离自然越来越远，持续发展的可能性越来越微。因此，从力所能及把居住的小环境景观设计做起，让其回复生气盎然的状态，进而投身大环境的改造与建设，把荒漠秃山瘠地流沙铺上植物、育成福地、为民所用，这是常系景观设计者心中的理想，也是圆我们中国梦心念的一环。

韦子刚 博士
GVL 怡境国际设计集团 主席
香港中华总商会会董、环境关注委员会主席、
英国约克大学环境生态学博士、英国政府环境部研究员

GREENVIEW LANDSCAPE INTERNATIONAL DESIGN GROUP(GVL)

Greenview Landscape International Design Group (GVL) is an Overseas Member of British Association of Landscape Industries (BALI) and a Corporate Member of the American Society of Landscape Architects (ASLA). GVL established head office in Guangzhou in 2000 and has set up Branches in Shenzhen,Changsha,Beijing, Shanghai, also Central South Eco-tourism Planning & Design Limited and PermaGreen Tourism Management Pty Ltd have subsequently opened.

Our main business is resource management and operations of scenic spots and specific tour, the landscape planning and design for star hotels, tourist resorts, commercial complexes, boutique residential communities, municipal projects and ecological projects. We are accomplished in environment planning, soft and hard landscape design, water feature design, water supply and drainage design, electrical and lighting design as well as the outdoor audio equipment detailing. We are well versed in all stages of the consultation and design process, as well as in project management, particularly in the area of design innovations for large-scale integrated residential developments. From GVL's business ideas, our approach has been: our reputation is based on the high standard design and our results are based on excellent service. With advanced and pragmatic design concepts, professional technical service, and world-class standards of project management, we have formed a whole service chain.

We have worked with many renowned developers, including VANKE, EVERGRANDE, POLY, WANDA, COLI and other top-tier real estate developers. Now our projects have spread all over China in more than 100 cities.

After more than a decade of innovation and high-quality service, GVL has won a great reputation in the landscape design industry and has been awarded for many times including Gold Award - Best Landscape Design Company at the Chinese Properties Economics Summit; Gold Brick Award - Best Design Company at the BOAO 21st Century Real Estate Forum; GVL has been nominated for 100 best professional companies in the China property sector; the top 10 landscape design companies in China; the top 10 landscape design of commercial & residential architectural design market in China.

怡境国际设计集团

怡境国际设计集团（Greenview Landscape International Design Group），简称GVL，是英国国家园景工业协会（BALI）海外会员、美国景观设计师协会（ASLA）企业会员。公司于2000年进驻广州设立总部，并先后设立深圳、长沙、北京、上海等区域公司、中南生态旅游规划设计有限公司及广东璞境旅游管理有限公司。

GVL的业务范畴主要包括旅游景区及专项旅游的资源管理运营；星级酒店、旅游度假区、商业及综合项目、精品居所、市政项目、生态旅游等的规划及景观设计等。在环境规划、园林设计、植物造景、水景设计、户外给排水、电气照明及背景音乐工程设计方面有着尤为丰富和深厚的实践经验，特别是在大型综合性开发项目上，能因地制宜地进行全面而细致的把控。秉承“设计创造品牌，服务雕琢精品”的经营理念，凭借先进的设计理念、过硬的专业技术及国际水准的现场跟踪，形成了一条完整的服务链。

GVL先后与众多知名地产商取得了合作，合作方包括万科地产、恒大地产、保利地产、万达集团、中海地产等顶级地产开发商，已完成和进行中的项目遍及全国100多个城市。

经过十多年来不断的设计创新及持续优质的后期服务，GVL的品牌得到了业内的肯定并屡获殊荣，其中包括中国地产经济主流峰会“最受推崇景观建筑设计公司”金鼎奖、博鳌21世纪房地产论坛“推动中国房地产发展卓越设计机构”金砖奖、中国房地产业最具合作价值专业公司100强、景观设计公司10强、中国民用建筑设计市场排名景观榜10强等。

Full (Overseas) Member of BALI (British Association of Landscape Industries)
英国国家园景工业协会（BALI）海外会员

Corporate Member of ASLA (American Society of Landscape Architects)
美国景观设计师协会（ASLA）企业会员

CONTENTS 目录

COMMERCIAL AREA	19	**都市新地标**
White Swan Hotel, Guangzhou	20	白天鹅宾馆，广州
ST.Regis Hotel, Changsha	34	瑞吉酒店，长沙
W Hotel, Changsha	40	W 酒店，长沙
Yunda Sheraton Hotel, Changsha	48	运达喜来登酒店，长沙
Zhonggeng Sheraton Hotel, Fuzhou	52	中庚喜来登酒店，福州
Yunda Central Plaza Commercial Area, Changsha	58	运达中央广场商业区，长沙
Zhonglong Jiaxi Center, Changsha	66	中隆嘉熙中心，长沙
Zhongtian the Future Ark, Guiyang	72	中天·未来方舟，贵阳
Zhongtian International Finance Center, Guiyang	80	中天·国际金融中心，贵阳
CULTURAL TOURISM+	84	**文化旅游 +**
Congdu International Conference Center, Guangzhou	86	从都国际会议中心，广州
Flower Town, Guangzhou	100	花山小镇，广州
Wanda Cultural Tourism City, Guangzhou	106	万达文化旅游城，广州

Poly Silver Beach, Yangjiang 108 保利银滩，阳江

Poly & Shunfeng Bunlos Secret Land, Yangjiang 120 保利顺峰·北洛秘境，阳江

Baodun Hot Spring Town, Yingde 128 宝墩湖温泉小镇，英德

Guangwu Cigar Town, Danzhou 134 广物·雪茄风情小镇，儋州

Poly Haitang Bay, Sanya 136 保利·海棠湾，三亚

Wanda Cultural Tourism City, Wuxi 140 万达文化旅游城，无锡

RESIDENTIAL AREA 144 现代精品居所

Poly Tangyue, Nanjing 146 保利·堂悦，南京

Poly Beautiful River Coast, Nanjing 156 保利·西江月，南京

Poly Grand Mansion, Fuzhou 164 保利·天悦，福州

Poly Westriver Imagination, Fuzhou 172 保利·西江林语，福州

Zhonggeng Premier Luxury River, Fuzhou 180 中庚·香江万里，福州

Poly City, Putian 186 保利城，莆田

Jinke Lounge Impression, Zhangjiagang 192 金科·廊桥印象，张家港

CONTENTS 目录

Everbright Top of the World, Dongguan 198 光大·天骄峰景，东莞

Four Seasons Garden, Dongguan 214 四季豪园，东莞

Everbright Beautiful Country, Dongguan 224 光大·锦绣山河，东莞

Vanke Dream Town · Ming, Guangzhou 238 万科城·明，广州

Agile Flowing Garden, Guangzhou 246 雅居乐·小院流溪，广州

Financial Street Rongsuihuafu, Guangzhou 252 金融街·融穗华府，广州

Yuexiu Starry Haizhu Bay, Guangzhou 258 越秀·星汇海珠湾展示区，广州

Poly i Cube, Guangzhou 264 保利 i 立方，广州

Poly Central Park, Heshan 270 保利中央公园，鹤山

The Orchidland, Huizhou 278 悠兰山，惠州

Yunda Central Plaza Residential Area, Changsha 288 运达中央广场居住区，长沙

Zhujiang Sun Town, Chongqing 296 珠江太阳城，重庆

Sincere Star Metropolis, Chongqing 300 协信星都会·溪山墅，重庆

Jinke City, Chongqing 304 金科城，重庆

Jinke Glamour Community, Xi'an	310	金科·天籁城，西安
Longfor Chianti International, Xi'an	314	龙湖·香醍国际，西安
Jinke Royal Spring Villa, Beijing	322	金科王府，北京
Dongsheng Dawn Garden, Zibo	328	东升·曦园，淄博
Jinke Sunshine Town, Qingdao	334	金科·阳光美镇，青岛
URBAN DEVELOPMENT	338	**城市开放空间**
Zhucheng City Square, Guiyang	340	筑城广场，贵阳
Poly Pazhou Eyes Greenland Park, Guangzhou	346	保利琶洲眼绿地公园，广州
Nansha Wetland Park, Guangzhou	348	南沙湿地公园，广州
Panlong Mountain Park, Queshan	352	盘龙山公园，确山
Daxiong Mountain Gaofeng Lake, Xinhua	356	大熊山高峰湖，新化
West Riverside Commercial Tourism Landscape Zone, Changsha	360	滨江新城西岸商业旅游景观带，长沙

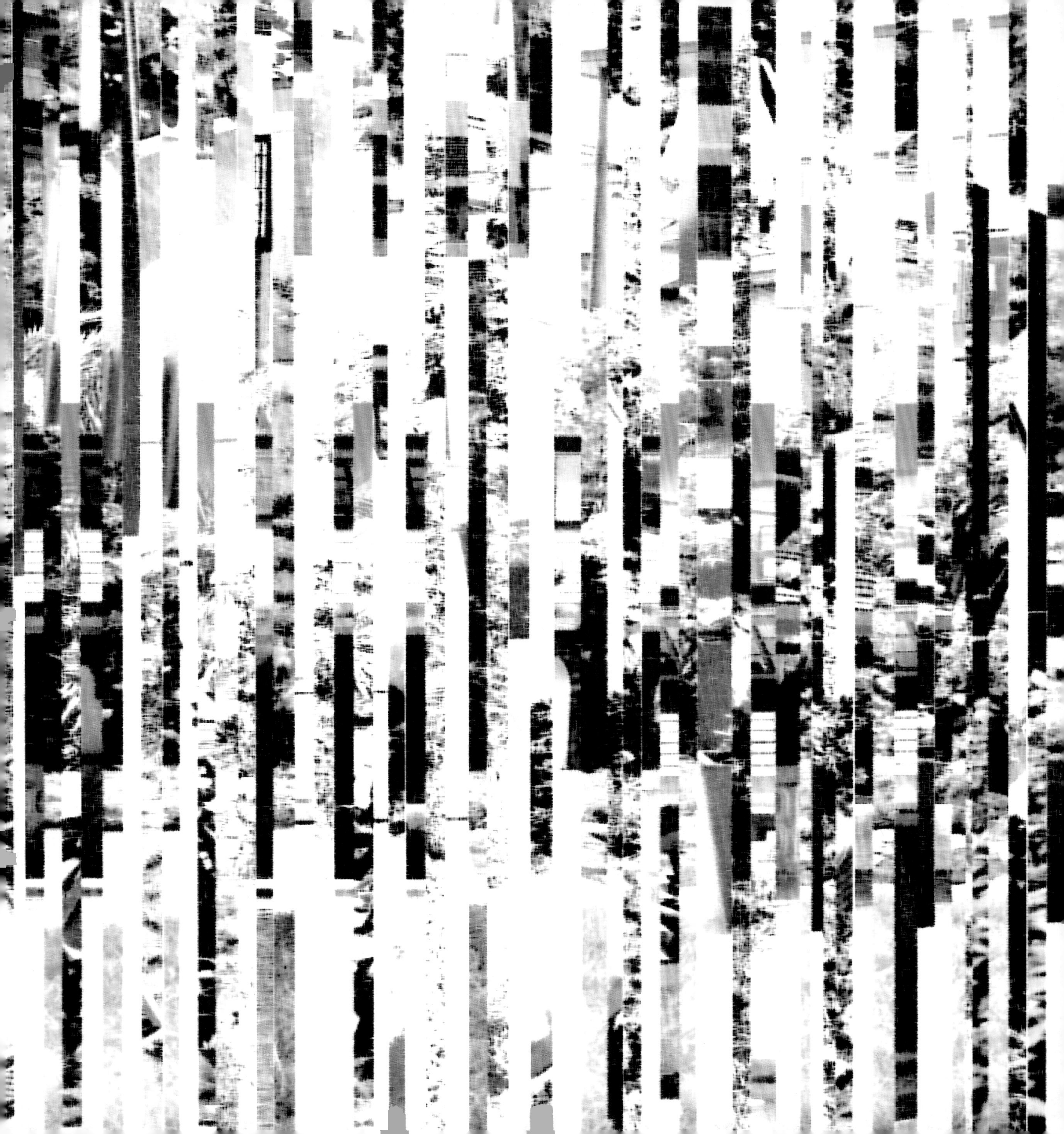

COMMERCIAL AREA

都市新地标

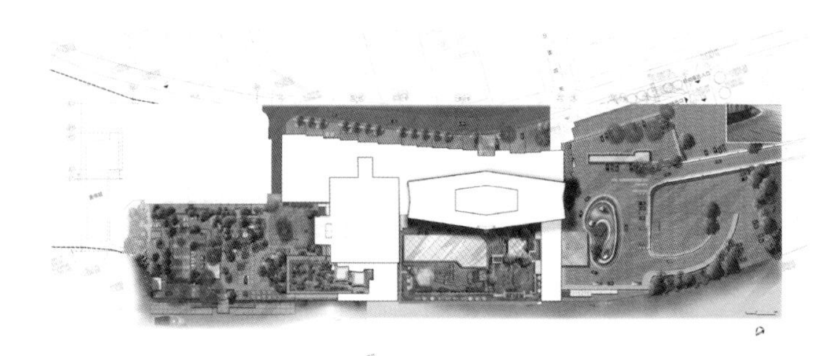

White Swan Hotel, Guangzhou
白天鹅宾馆，广州
——白天鹅的华丽转身，新岭南的完美诠释

Developer: White Swan Hotel
Project Type: Five-star Hotel
Project Area: 30,000 m²
Design Content: Landscape Design of the Hotel
Design Period: 2013

委托单位：白天鹅宾馆
项目类型：五星级酒店
项目面积：30 000 m²
设计内容：酒店景观设计
设计时间：2013 年

The White Swan Hotel is located in the south of Shamian Island, which is a heaven of tranquility amid the hustle and bustle of the city, majestically overlooking the mother river of Guangzhou —— the Pearl River. Opened in 1983, White Swan Hotel was jointly developed by Mr. Henry Fok Ying Tung and Guangdong government. It is the first five-star hotel in mainland China that demonstrates the success of the reform and opening-up policy. It also ranks first in the country on many other aspects, making itself well-known both at home and abroad. Because of its long history and cultural value, it becomes an important historical mark deeply rooted in the heart of local citizens.

As the world's major hotel brands are entering in China's high-end hotel market, the White Swan Hotel is facing great challenges. Standing at a new historical starting point, to break through, innovate, reform and continue to lead the high-end starred hotel market in Guangzhou is the purpose of our design.

As a classic landscaping work of Lingnan style, the White Swan Hotel has successfully combined the local cultural features with the spirit of the times, creating a leisurely atmosphere without losing the elegant cultural connotations.

The design has retained White Swan Hotel's classic characteristics and increased some high-end modern elements to upgrade the leisurely and elegant environment, to present the charm of traditional Lingnan style and to keep pace with the modern society. The renovated hotel has met the people's requirements for high-grade enjoyment, and the once charming "swan" has regained its beauty and elegance!

白天鹅宾馆坐落于广州闹市中的“世外桃源”——沙面岛南边，屹立于母亲河——珠江之畔。白天鹅宾馆于1983年建成开业，是由霍英东先生与广东省政府投资合作兴建的，被誉为印证改革开放成功典范的，国内由中国人自主管理的首家高星级酒店。作为中国首家五星级宾馆，白天鹅宾馆创造了一个又一个的“全国第一”，蜚声海内外。其历史文化价值，在广州本地更是深入人心，成为市民心中的一个重要历史标志。

随着世界各大著名品牌酒店纷纷抢滩中国高端酒店市场，白天鹅也面临着新的巨大挑战。站在新的历史起点上，力求突破，创新变革，再创辉煌，继续引领广州高端星级酒店市场是本次设计的宗旨。

作为广州地区酒店岭南园林景观的经典之作，白天鹅宾馆将地域文化特色与时代精神完美融合，在创造一种从容不迫境界的同时又不乏优雅的文化内涵。

本次设计在保留白天鹅传统经典特色的基础上，与时俱进，全新演绎新时代的审美情怀，在白鹅潭从容、优雅的环境中，融入现代独特的设计元素，既呈现了传统的岭南风韵，又展示了现代时尚的一面，满足现代人对生活追求极致享受的需求，让曾经风华绝代的那只“天鹅”更加光彩照人，经典永恒！

main entrance of the new White Swan Hotel in the night, unpretentious but distinguished　夜幕中的新白天鹅主入口，低调而尊贵

白天鹅宾馆，广州

1 | 2
  | 3

1 reflection pool at the main entrance
2 pool and artificial hills
3 sub entrance of White Swan Hotel and the planting beds

1 主入口倒影池
2 一池三山
3 白天鹅宾馆次入口与具有盆景理念的种植池

白天鹅宾馆，广州

1	4
2 | 3	5

1 high and low planting beds at the entrance
2 delicate landscape wall by the swimming pool
3 corridor in the peaceful garden
4 green wall along the pool
5 sitting on the bank of the Pearl River and enjoying great river views

1 入口前高低起伏的种植池
2 细致入微的泳池区景观墙
3 廊架外，园境幽静
4 水池边，绿墙尽显生态之美
5 依水而栖，享一线江景

白天鹅宾馆，广州

1	2	5
3	4	

1-2 railings of Lingnan Style
3 details of waterscape
4 luxuriant vegetation on the green wall
5 quiet and beautiful courtyard in the night

1–2 充满岭南意象的栏杆
3 水景细部
4 绿墙上的葳蕤草木
5 夜景里清幽的庭院

1	2
	3	4

1 hometown water for generations
2 shimmering swimming pool in the night
3 bird's-eye view of the backyard of White Swan Hotel
4 swimming pool is near to the Pearl River

1 几代岭南人的故乡水
2 夜色中的泳池灯光闪烁
3 鸟瞰白天鹅后场花园
4 泳池与江面相邻

White Swan Hotel, Guangzhou

白天鹅宾馆，广州

1 | 2 | 3
  | 4 | 5

1 tropical scenery around the swimming pool
2 details of the terrace by the swimming pool
3 lighting design of the swimming pool
4-5 swimming pool in the night

1 泳池边，一派热带好风光
2 泳池边平台细部
3 泳池灯光
4-5 泳池夜景

1	2	5
3	4	

1 outdoor terrace in the night
2 green wall and lattice window, collision between pioneer and tradition
3 modern and elegant pavilion
4 water pool in the night
5 a corner of the outdoor courtyard

1 户外平台夜景
2 绿墙与花窗，先锋与传统的碰撞
3 亭子造型前卫空灵
4 水池夜景
5 户外庭院一角

HOTEL
ST.REGIS

ST. Regis Hotel, Changsha
瑞吉酒店，长沙
——国际风范，高尚格调

Developer: Hu'nan Yunda Real Estate Development Co.,Ltd.
Project Type: Five-star Hotel
Project Area: 14,980 m^2
Design Content: Landscape Conceptual Design & Production Design
Design Period: 2010

委托单位：湖南运达房地产开发有限公司
项目类型：五星级酒店
项目面积：14 980 m^2
设计内容：景观方案及施工图设计
设计时间：2010 年

Regis Hotel is ideally located in the core area of WuGuang Business Circle. The design holds the idea of "low carbon and environmentally friendly" and takes some energy-efficient and intelligent measures, trying to build a paradigm of star hotel and a highland of fashion life in the city of Changsha. Based on high-standard design principles and with emphasis on the application of new ideas, new materials and new technologies, the landscape design has fully considered the relation with the surroundings to create reasonable functions and comfortable spaces, and to keep the landscapes in harmony with the buildings and the surrounding environment.

瑞吉酒店地处武广商圈核心地段。设计秉承“低碳环保”理念，运用节能、智能元素，使项目成为长沙未来持续领先的星级酒店典范及时尚生活高地。景观设计以高标准的设计原则，注重新理念、新材料、新技术的应用，考虑与周围环境的协调、呼应，设计功能合理，空间舒适，达到建筑与景观的有机统一。

ST. Regis Hotel, Changsha

exterior of the waterscape at the entrance of St. Regis Hotel　瑞吉酒店入口水景外景

1 | 3
2 | 4

1 magnificent waterscape at the entrance
2 natural plank road
3 cascade and plank road
4 flowing water and rough stones

1 入口飘绢水景大气磅礴
2 自然木平台登道
3 叠水与木栈道相互呼应
4 流水与原石的相融相衬

ST REGIS

1 | 4
2 | 3 | 5

1 night view of the waterscape at the main entrance of St. Regis Hotel
2 gurgling waterfalls and calm pool
3 natural landscape rocks
4 night view of the cascade
5 night view of the waterscape (interior)

1 瑞吉酒店正门水景夜景
2 潺潺瀑布叠水与平静水面形成强烈对比的美感
3 自然山水石
4 叠水夜景
5 内侧水景夜景

HOTELS

W Hotel, Changsha
W 酒店，长沙
——艺术与视觉盛宴

Developer: Hu'nan Yunda Real Estate Development Co.,Ltd.
Project Type: Five-star Hotel
Project Area: 18,500 m²
Design Content: Landscape Conceptual Design & Production Design
Design Period: 2010

委托单位：湖南运达房地产开发有限公司
项目类型：五星级酒店
项目面积：18 500 m²
设计内容：景观方案及施工图设计
设计时间：2010 年

W Hotel, Changsha is located in the heart of WuGuang Business Circle, a district where the traffic network is dense. Convenient traffic will quickly gather people, goods and wealth, and lowland effect will appear soon.

The design holds the idea of "low carbon and environmentally friendly" and takes some energy-efficient and intelligent measures, trying to build a paradigm of star hotel that interprets top-quality lifestyle. Based on high-standard design principles and with emphasis on using new ideas, new materials and new technologies, the landscape design has fully considered the relation with the surroundings to create reasonable functions and comfortable spaces, and to keep the landscapes in harmony with the buildings and the surrounding environment.

W 酒店地处武广商圈核心地段，密布城市交通网，便利的交通将迅速聚集人流、物流、财流，洼地效应很快显现。

设计秉承"低碳环保"理念，运用节能、智能元素，使项目成为长沙未来持续领先的星级酒店典范及时尚生活高地。景观设计以高标准的设计原则，注重新理念、新材料、新技术的应用，考虑与周围环境的协调、呼应，设计功能合理，空间舒适，达到建筑与景观的有机统一。

waterscape at the entrance of W Hotel　W 酒店入口水景

1	4
2 | 3	5

1 jumping fountains at the main entrance
2 parking entrance
3 colorful vegetation zone in the light
4 waterscape and green landscape at the entrance of the commercial level
5 floor plan of the hotel entrance

1 跳泉给主入口带来动感
2 酒店停车场入口
3 灯光下色彩丰富的植物带
4 商业入口层级水景结合绿化
5 酒店入口平面图

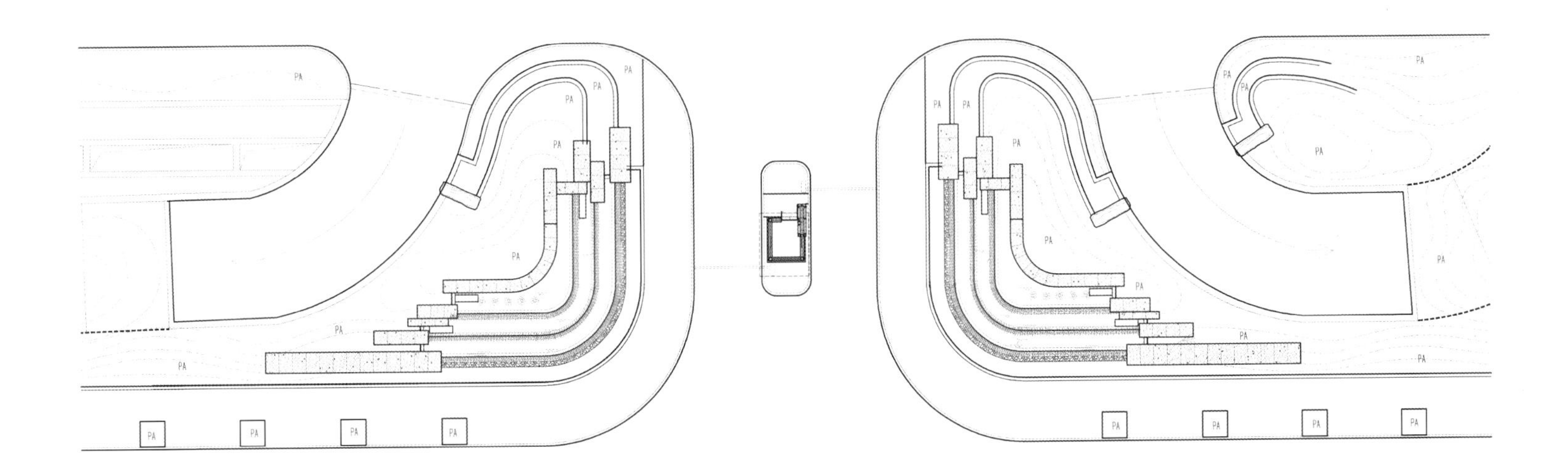

1 | 4
2 | 3 | 5

1 waterscape in the roof garden
2 path in the roof garden
3 landscape wall in the roof garden
4 tearoom in the roof garden
5 waterscape at the entrance of the roof garden

1 天台花园水景
2 天台花园特色园路
3 天台花园景墙
4 酒店天台花园茶室
5 天台花园入口水景

	1
2	4
3	5

1 site plan
2 section 1 of the roof garden of W Hotel
3 section 2 of the roof garden of W Hotel
4 section 1 of the roof garden of St. Regis Hotel
5 section 2 of the roof garden of St. Regis Hotel

1 总平面图
2 W 酒店天台花园剖面图一
3 W 酒店天台花园剖面图二
4 瑞吉酒店天台花园剖面图一
5 瑞吉酒店天台花园剖面图二

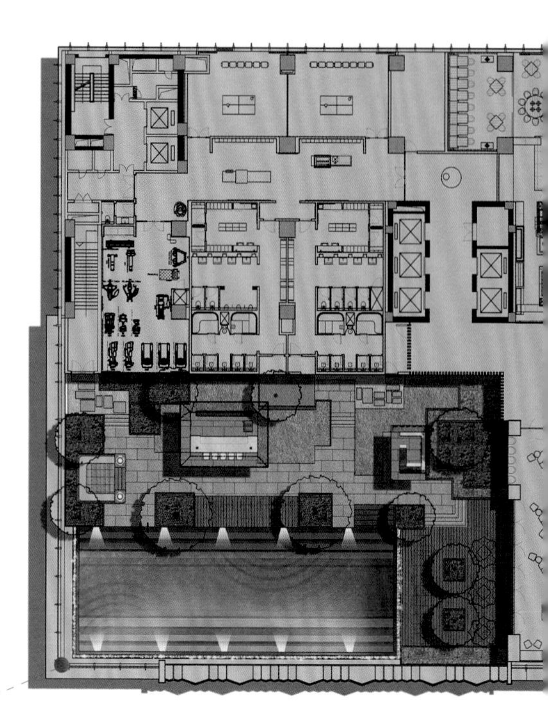

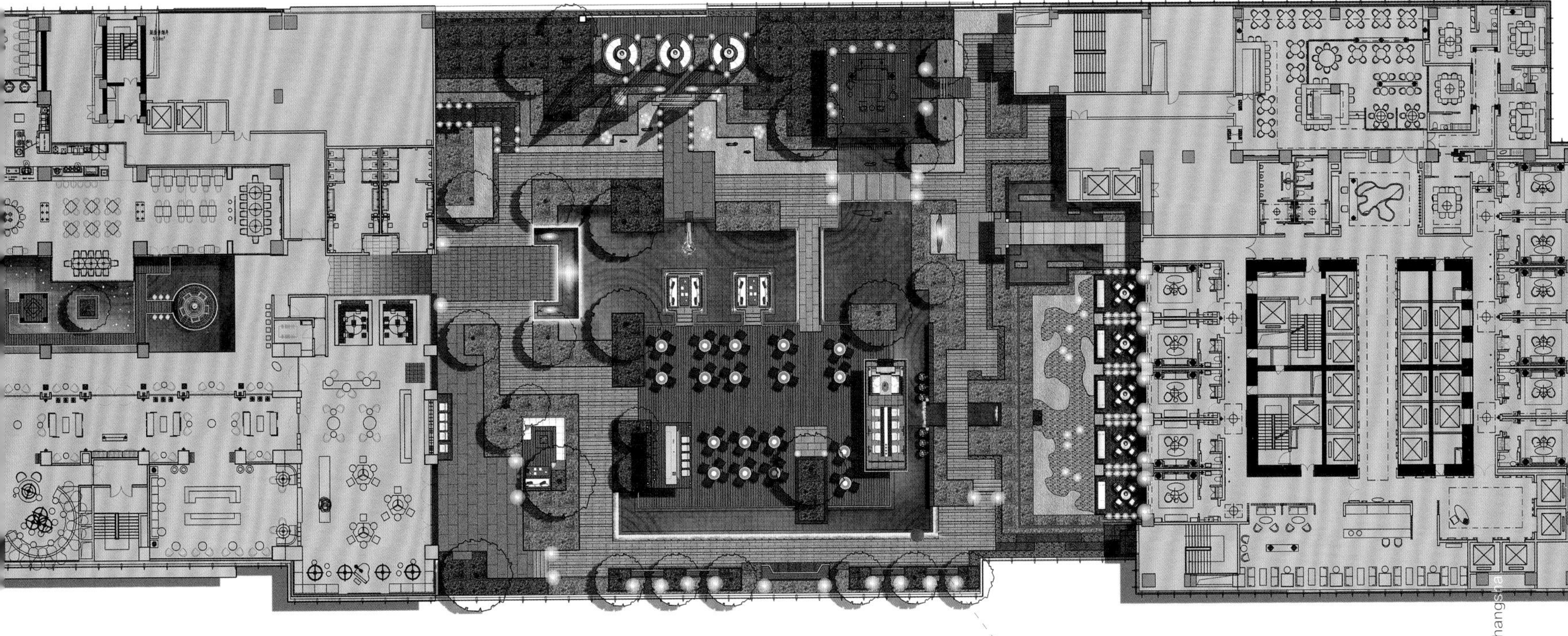

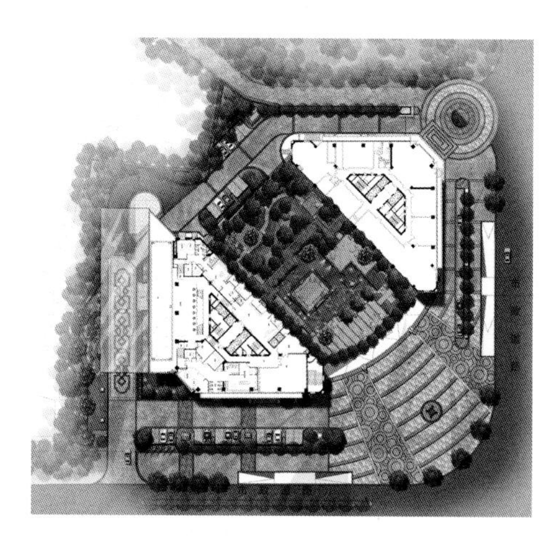

Yunda Sheraton Hotel, Changsha
运达喜来登酒店，长沙
——浪漫诗意花园，商务休闲情调

Developer: Hu'nan Yunda Real Estate Development Co.,Ltd.
Project Type: Five-star Hotel
Project Area: 26,000 m^2
Design Content: Landscape Conceptual Design & Production Design
Design Period: 2006-2007

委托单位：湖南运达房地产开发有限公司
项目类型：五星级酒店
项目面积：26 000 m^2
设计内容：景观方案及施工图设计
设计时间：2006—2007 年

Landscape design of this project has won many awards, including the International Award (Principal Award Winner) issued by BALI in 2008; the 7th Golden Pillow Award "The Best CBD Business Hotel in Mid-South China"; "2009 the Best Starwood Hotel in Asia Pacific"; "2009-2010 China's Most Charming Business Hotel" by the 10th China Hotel Golden Horse Award; "2010 Top 100 Hotels" by China Travel Award; "The Best Business Hotel" in 2010 Charming Hotel List.

该项目 2008 年获英国国家园景工业协会（BALI）授予的景观设计国际项目金奖；第七届金枕头奖之"中国中南部地区最佳 CBD 商务酒店"；"喜达屋亚太区 2009 年度最佳酒店"；第十届"中国饭店金马奖"之"2009 – 2010 中国最佳商务酒店"；2010 中国旅行奖之"百佳酒店"；2010 酒店魅力榜之"最佳商务酒店"。

Yunda Sheraton Hotel is the north tower of Yunda International Plaza, which is located at Songguiyuan core section of Changsha CBD. The style of landscape architecture, furniture and pavement design in the first floor echoes with the business atmosphere of the building facade, meanwhile, terraced waterscape and plants make it more natural and flexible. The overall layout of the roof top garden proceeds with flowing lines and concise straight lines. Landscape elements are situated properly, such as waterscapes, waterfront wooden platform, etc. Simple and pure corridor and landscape framework keep in line with contemporary aesthetic taste and add a breath of resort-style life as well.

长沙运达喜来登酒店是运达国际广场的北塔建筑，位于长沙市商业中心区的松桂园核心地段。酒店首层的景观建筑、小品与铺装设计的风格均与建筑外立面简约的商务气质相呼应，并用叠级水景和植物带来自然灵动的气氛。天台花园的整体布局以流畅的线条和简练的直线进行构图，景观元素主从分明、关系明确，并设置了水景和亲水木平台，在廊架和景观构架等景观单体和小品设计上，呈现出简单、纯粹的特点，符合现代人的审美品位，同时为酒店增添了度假式的生活气息。

1
2 | 3

1 dynamic waterscape at the hotel entrance
2 landscape at the hotel entrance
3 waterscape on the ground floor

1 酒店入口处富于变化的水景
2 酒店入口景观
3 首层广场水景

1 | 3
2 |

1 roof garden of holiday style is perfect for wedding ceremony
2 hand drawing of koi pond
3 waterfront platform, tree pool in water and cascade wall with smooth and elegant straight lines

1 度假式的生活气息，让天台花园成为举办婚礼的绝佳场所
2 锦鲤池手绘效果图
3 亲水平台、水中树池、落水景墙均为简洁流畅的直线构图

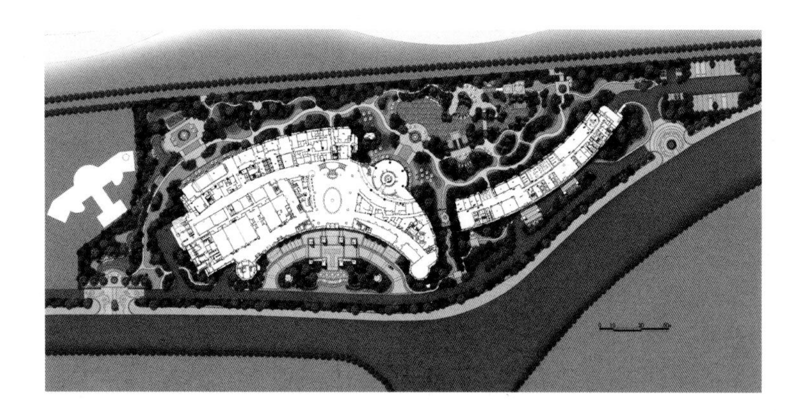

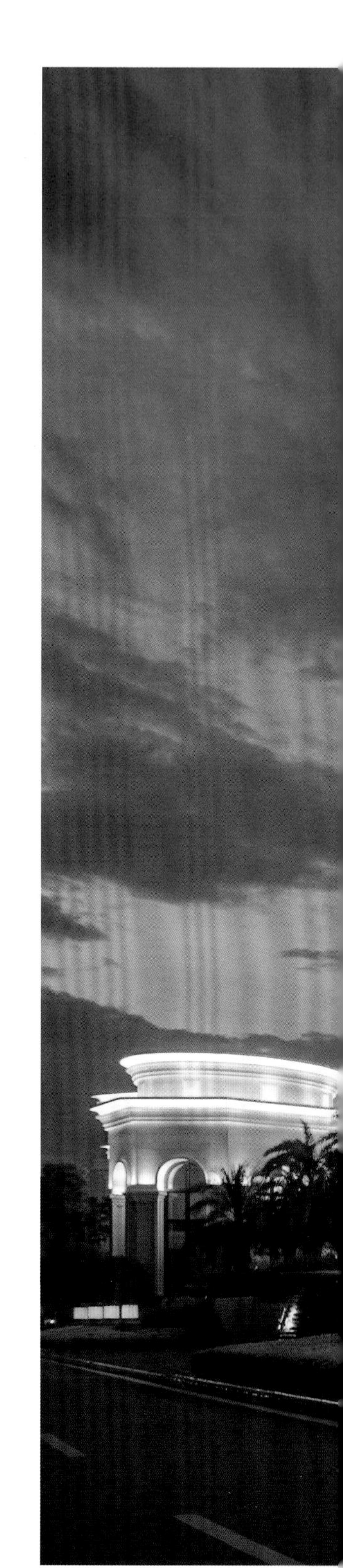

Zhonggeng Sheraton Hotel, Fuzhou
中庚喜来登酒店，福州
——赏闽江之景，享法式情怀

Developer: Fujian Zhonggeng Real Estate Co., Ltd.
Project Type: Five-star Hotel
Project Area: 77,337 m^2
Design Content: Landscape Conceptual Design & Production Design
Design Period: 2011

委托单位：福建中庚置业有限公司
项目类型：五星级酒店
项目面积：77 337 m^2
设计内容：景观方案及施工图设计
设计时间：2011 年

Zhonggeng Sheraton Hotel is located in the east side of New Town Convention & Exhibition Island and on the south bank of Mingjiang River in Cangshan District, Fuzhou. It is a super-platinum five-star resort hotel that appears as paired towers, and favorable enough for commercial activities, conferences, leisure and entertainment, etc.

There is a certain height difference between the south of the project and urban road, forming a genuine feeling of territory. The hotel building is in neo-classical style and the landscape design extends and strengthens the luxury architectural style especially in the entrance area. Designers pay equal attention to grace and classical details, reflecting the star-rated quality by the detail design of monomer, furniture and pavement. And they make the most of existing resources, using height difference to form terraced greening. Besides, outer river views interplay with the inner courtyard. Distinguished terraced flower bed spreads in accordance with the site typography and incorporates with the site and the building naturally, which fully demonstrates the design skills. In addition, a relaxed and ecological holiday atmosphere is created thanks to the green plants, sculpture and fabric.

中庚喜来登酒店位于福州市仓山区东部新城会展岛东侧、闽江南岸。酒店整体定位为超白金五星级度假酒店，集商务活动、会议、休闲、娱乐于一体，呈双塔式建筑布局。

项目南面与市政道路有一定的高差，形成了天然的领域感。酒店建筑为新古典风格，景观设计延续并强化了建筑风格所特有的奢华感，在入口处强化新古典的厚重奢华感。大气度与经典细节并重，在小品、铺装、单体等细部设计上体现星级品质，并充分利用场地现有资源，如利用高差形成层级绿化，因借外部江景形成外江内庭的格局，顺势而成的层级花坛尊贵大气，与场地、建筑自然融合，充分展示设计功底，同时，在绿植、雕塑、布品上营造出生态轻松的度假氛围。

entrance of the hotel　酒店入口

1 night view of the five-star car entrance
2 night view of the landscape wall at the sub entrance
3 european-style landscape lamps
4 landscapes along the municipal road

1 五星级车行入口夜景
2 次入口景墙夜景
3 欧式景观灯
4 市政车行道景观

1
2 | 3

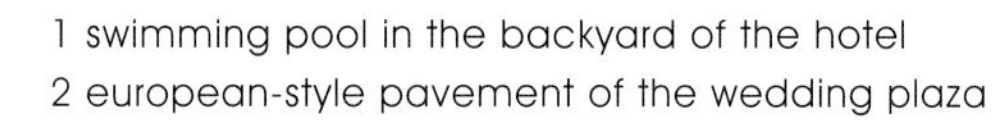

1 swimming pool in the backyard of the hotel
2 european-style pavement of the wedding plaza
3 quiet and comfortable garden path in the night

1 酒店后场泳池景观
2 婚庆广场欧式铺地更显尊贵
3 园路夜景安静且舒适

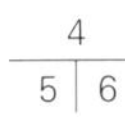

4 comfortable wooden platform by the swimming pool
5 water bar pavilion
6 the swimming pool is the focus of the hotel garden

4 泳池边木平台高端舒适
5 水吧亭是景观与功能为一体的构筑物
6 泳池景观是五星酒店全园的焦点

Yunda Central Plaza Commercial Area, Changsha
运达中央广场商业区，长沙
——时尚生活新高地

Developer: Hu'nan Yunda Real Estate Development Co.,Ltd.
Project Type: Commercial Area
Project Area: 23,600 m^2
Design Content: Landscape Conceptual Design & Production Design
Design Period: 2010

委托单位：湖南运达房地产开发有限公司
项目类型：商业空间
项目面积：23 600 m^2
设计内容：景观方案及施工图设计
设计时间：2010 年

Ideally located in the heart of Wuguang Business Circle, Yunda Central Plaza sits at the northwest corner of the intersection where Changsha Road and Shawan Road meet. Accommodating two super-platinum five-star hotels, 5A offices, international brand shopping mall and commercial street, international served apartment and the 6th generation diamond residences, Yunda Central Plaza, without any doubt, will become the top commercial complex in Changsha and even in central and south China.

The landscape design is inspired by the idea of "low carbon" to integrate European style with Chinese style. By employing water features, landscape walls, lamps and green plants, Yunda Central Plaza will be the ecological community paradigm that leads the fashion lifestyle of Changsha.

panorama of Yunda Central Plaza 运达中央广场全景

ST. REGIS
运达中央广场
YUNDA CENTRAL PLAZA
华中地区商业翘楚
10万平方米国际名品购物中心
Commercial leader in central China
100,000 square meters international brands shopping center
运达中央广场
YUNDA CENTRAL PLAZA

运达中央广场位于长沙大道与沙湾路相交的西北角，是武广商圈核心地段，交通便利。国际白金五星级酒店、国际5A写字楼、国际名品商场、国际名品步行街、国际酒店式公寓、中国第六代钻石华宅组成的六大完美业态组合，必然让项目成为长沙乃至中南地区顶尖的商业综合体。

景观设计秉承“低碳环保”理念，融欧陆风情和中式风格为一体，运用水景、景墙及灯光系统等造景元素，结合生态自然的园林绿化，使项目成为长沙未来持续领先的生态社区典范及时尚生活高地。

1 | 2

1 modern and elegant commercial street
2 entrance landscape of the commercial street

1 现代简洁的商业内街
2 商业内街入口景观

运达中央广场商业区，长沙

1	4
2 | 3	5

1 exquisite fountains
2 Phoenix sylvestris trees
3 tree pool and seats
4 tree pool and pavement in harmony
5 commercial street is amazing at dusk

1 精致的涌泉水景，又是功能性的室外活动展台
2 银海枣树阵，高贵且风情
3 树池结合座椅，为商业人群提供休憩场地
4 陈列的树池美景与铺地的和谐统一
5 待入夜的商业街，更显魅影迷人

运达中央广场商业区，长沙

1	2	3
4	5	6

1 exquisitely designed wooden tree pool
2 hollowed-out stainless steel
3-6 details of the tree pool

1 木质树池精致且有细节
2 不锈钢镂空细致处理
3-6 树池细节

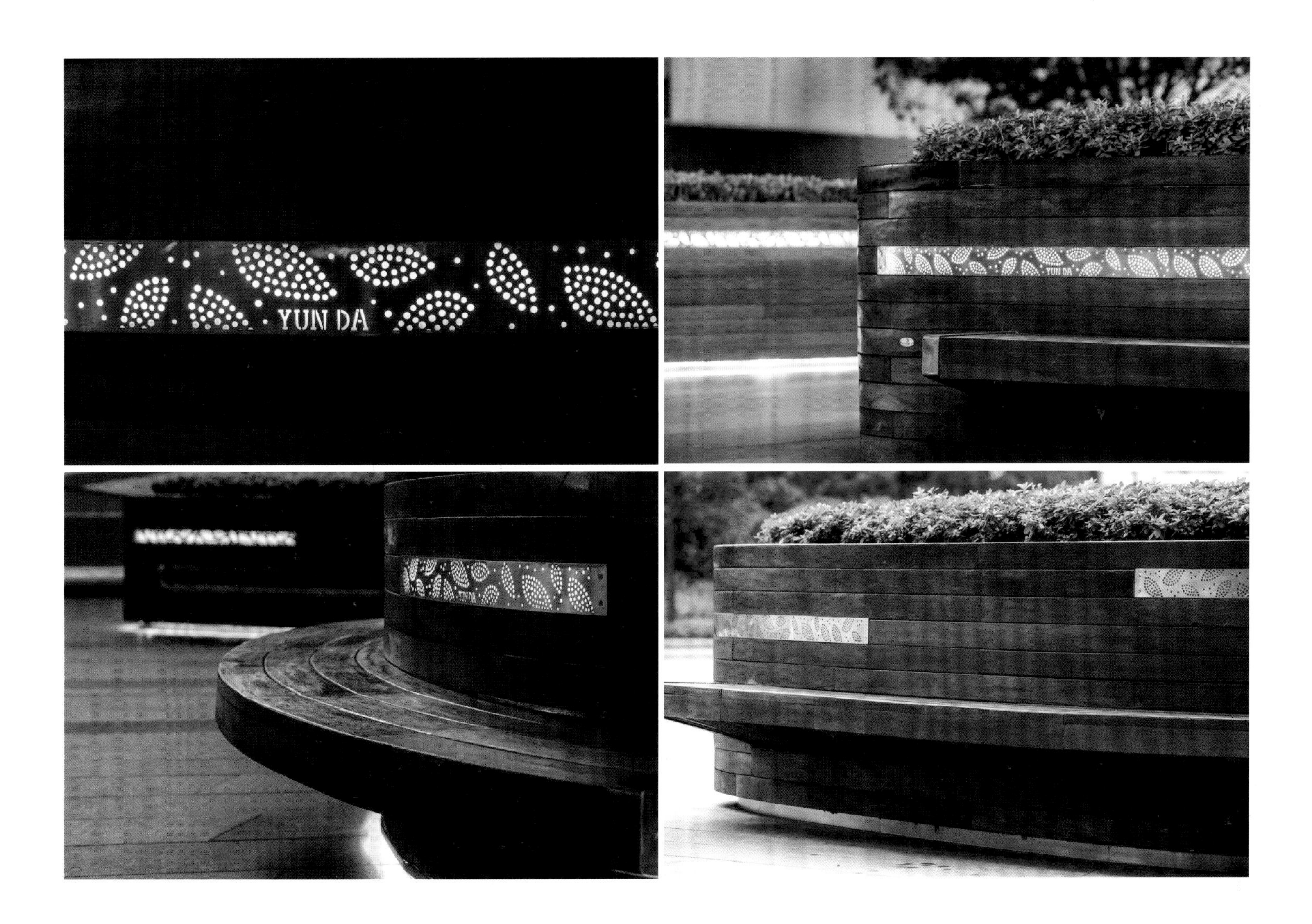

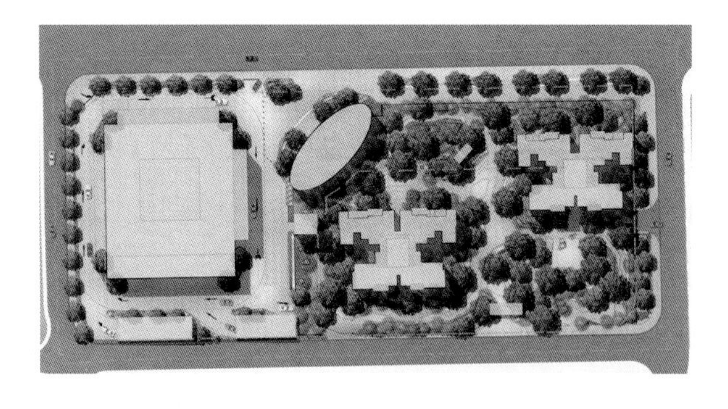

Zhonglong Jiaxi Center, Changsha
中隆嘉熙中心，长沙
——高端城市商务综合体

Developer: Hu'nan Jiaxi International Real Estate Development Co.,Ltd.
Project Type: Commercial Area
Project Area: 21,138 m^2
Design Content: Landscape Conceptual Design & Production Design
Design Period: 2013

委托单位：湖南省嘉熙国际房地产开发有限公司
项目类型：商业空间
项目面积：21 138 m^2
设计内容：景观方案及施工图设计
设计时间：2013 年

The project is ideally located on Renmin Middle Road of Changsha City, surrounded by complete supporting facilities. It is a typical example of high-end, luxury and unpretentious urban complexes of new Asian style. Designed with the idea of "wonderful time, delicate life and beautiful memories", and taking advantage of the altitude difference, it has created a series of dynamic spaces: the extraverted commercial spaces and the introverted office spaces. Multi-level landscape system composed of vegetations, green walls and cascades has built a tranquil environment where people will slow down and experience the passage of time in peace.

It has paid much attention to the design of nodes and details. Modern geometrical flowerpots, steps and seats as well as the pavements, flowers, grass and trees all combine together and create a multi-level landscape space.

项目位于长沙市人民中路，地理位置优越，周边配套设施完备，是新亚洲风格的典型代表，是新型高端低奢的代言。项目以"曼妙时光、精致人生、朝花夕拾"为设计理念，利用场地原有高差分隔出商业的外向与组团的内含两种不同的空间，增加空间的灵动性。层级绿化、垂直绿墙以及悬挂如链的跌水，不仅丰富了景观的层次，更营造了一种隔绝尘世喧嚣的环境，让奔波忙碌的人们放慢脚步，在芳影叠翠间静静感受时光流转。

项目对于节点与细部的设计用心考究，现代几何造型元素的花基、阶梯、座椅等景观语言，加上极具时代特色的铺装材料，将景观氛围层层渲染。在植物配置中，使用多层次的植物造景，选用大冠幅乔木结合错落有致的花基，围合出多层次的景观空间。

waterscape matches the square in front of the sales center, looking elegant and unpretentious　销售中心前广场与水景相辉映，低调与雅致尽显

1 | 2
3

1 delicate pavement of the square
2 planting pools and the entrance
3 multi-level waterscape outside the sales center

1 精致前卫的广场铺地
2 入口旁的种植池
3 销售中心外水景的多层次造景

4
5

4 pool and fountains at the main entrance of the sales center
5 pavement of the square in the light

4 销售中心主入口的水池喷泉，生机无限
5 灯光掩映下的广场铺地更显特色

1	2	5
3	4	

1 dialogue between soft landscape and hard landscape
2 skillful space transition
3 pleasant rest area and reflection pool in the garden
4 exquisite details of finishing
5 sales center in the night

1 软硬景的对话
2 巧妙的空间转换
3 园区怡人的休息区与镜水面
4 活泼而极致的饰面细节
5 夜幕下的销售中心

Zhonglong Jiaxi Center, Changsha

Zhongtian the Future Ark, Guiyang
中天·未来方舟，贵阳
—— 西部未来之城，山水国际都会

Developer: Zhongtian Urban Development Group
Project Type: Large-Scale Satellite Town
Project Area: 10,430,000 m^2
Design Content: Landscape Conceptual Design & Production Design
Design Period: 2011 as yet

The project has won the "Annual Media Recommended Pre-selling Housing Development" of the 10th Kinpan Award (West China) in 2015

Located in Yu'an. Anjing area of Yunyan district, the eastern old town of Guiyang, Future Ark enjoys the superior natural resources that provided by the mother river of Guiyang, a meandering 7 km river bank, which gives Future Ark fresh life and vigor.

The site of the project covers an area of 10,430,000 m^2 which is split into three sections such as tourism, high-end habitation and ecological landscape, including large-scale theme park, five-star spa resort hotel, comprehensive nursing home, theater, indoor ski field, marine museum, large shopping mall, large supermarket, etc. And the four symbolic high-rise buildings will house the world's top business, commerce, hotel and numerous world top 500 enterprises, which can be said to be the large -scale satellite town in Guiyang.

The overall project is positioned as the "Central Park Area", a collection of nature, living, office, tourism and other world-class facilities that implants planning concepts and lifestyle of international garden residence, and innovates the city living environment in Guiyang. After analyzing the residential and commercial system, landscape designers propose an overall landscape layout incorporating "mountain, water, wood and city". "Mountain" refers to create mountain park and high-end mountain villa landscape by making use of "Mo Jia Jian Mountain" in this area. "Water" means to create a 7 km-long riverfront leisure zone, i.e., the water of vigor, water of flexibility, water of spring. "Wood" means to retain and optimize the original mountain vegetation to construct ecological landscape for high-rise building area. "City" means to integrate modern and fashion with local context organically, highlighting the concept of the "city of the future, city of fashion and city of charm", shaping "the further city parlor".

委托单位：中天城投集团
项目类型：超大型卫星城
项目面积：10 430 000 m^2
设计内容：景观方案及施工图设计
设计时间：2011 年至今

该项目获得 2015 年度第十届金盘奖·西南西北赛区“预售楼盘类年度媒体推荐奖”

本项目位于贵阳市老城区东部的云岩区渔安安井片区，坐拥贵阳母亲河——南明河，两岸优越的自然资源，蜿蜒 7km 的天然河岸，为项目注入鲜活的生态命脉。

项目总占地面积达 10 430 000 m^2，分为旅游、高端人居、生态景观三大板块，包括大型主题公园、五星级温泉度假酒店、综合疗养院、大剧场、室内滑雪场、海洋馆、大型商业 MALL、大型超市等功能配套，四大门户超高层地标，将容纳世界顶级的商务、商业、酒店以及众多世界 500 强企业，堪称贵阳市内超大型卫星城。

项目整体定位为“中央公园区”，汇集了自然、生活、办公、旅游等世界顶级配套，将国际高尚公园住区的规划理念、生活方式植入这片土地，全面革新大贵阳的城居环境。景观设计结合对项目商住系统的分析，提出“山、水、林、城”的整体景观布局建议。以“山”为题，着力打造场地内原有的“莫家尖山”山体公园及山居高端别墅景观；“水”以南明河为纽带打造 7km 的沿河滨水休闲带，即活力之水、灵动之水、思源之水；“林”保留及优化场地内原有山体植被——营造林居生态高层区景观；“城”有机地将现代时尚动感之都融入当地文脉，凸显“未来之城”“时尚之城”“魅力之城”的概念，塑造“未来城市客厅”。

Planning site plan 规划总平面图

中天·未来方舟，贵阳

1 | 2

1 rendering of Nanming River right bank night view
2 rendering of mountain top road and landscape bridge night view

1 南明河右岸夜景效果图
2 山顶区道路及景观桥夜景效果图

POLLOK

1 rendering of commercial street Landscape
2 rendering of yacht club
3 rendering of commercial district and riverside night view

1 商业区街景效果图
2 游艇会所效果图
3 商业区及沿河夜景效果图

1 rendering of sales center square
2 rendering of Yansun Park on the mountain top night view
3 rendering of symbolic beacon on the mountain top night view

1 售楼部广场效果图
2 山顶区燕隼公园夜景效果图
3 山顶区标志性灯塔夜景效果图

COMMERCIAL AREA 都市新地标

Zhongtian International Finance Center, Guiyang
中天 · 国际金融中心，贵阳
—— 顶级商业综合体，高端金融创意地标

Developer: Zhongtian Urban Development Group
Project Type: Commercial Area
Project Area: 252,284 m^2
Design Content: Landscape Conceptual Design & Production Design
Design Period: 2014

委托单位：中天城投集团
项目类型：商业空间
项目面积：252 284 m^2
设计内容：景观方案及施工图设计
设计时间：2014 年

Located in the core area of Guiyang's CBD, Zhongtian International Finance Center houses high-end commercial spaces and well-decorated office spaces. The landscape is designed with the idea of "low carbon". By skillfully employing water features, landscape walls, landscape lamps and natural vegetation, it has created a landmark project of commerce, office and modern lifestyle in southwest China.

Based on advanced urban planning idea, the project is well planned to adapt to the local climate. With clever cost control and scientific eco strategy, it has created a modern, international and people-oriented financial center. Moreover, roof gardens with green spaces, F&B spaces, shops and art gallery are designed to connect all towers together. The designers have also taken the traffic flow into consideration: the roadway is set around the development which is separated from the footways to avoid noises and interference. In this way, it ensures a safe, quiet and comfortable working environment. In addition, the large green area creates an ecological recreation area, while the landscapes in and around the twin office towers greatly upgrade the environment of the entire region. And the world-renowned Marriott Hotel is the biggest highlight of this project. In the future, Zhongtian International Finance Center will play an important role in further development of Guiyang city.

项目地处贵阳重要商圈核心地段，由高端商业组合及高档精装修高层商业办公空间构成，景观设计秉承"低碳环保"理念，融现代商业空间为一体，运用水景、景墙及灯光系统等造景元素，结合生态自然的园林绿化，使项目成为西南地区持续领先的商业办公典范及时尚生活高地。

中天金融城以先进的城市规划理念为指导，合理整合资源、科学规划。针对地区气候特点，在适宜的成本控制下，融合了富有地域特色的节能生态策略，创造人性化的空间，充分体现了以人为本，生态环境为主的生态型、人性化、现代化、国际化的金融办公区特性。同时，设计师团队还通过屋顶花园的设置，为地块添加立体化的绿色休憩空间，并结合屋顶花园设置小型餐饮、商业、艺术展览，使屋顶成为联系各塔楼的文化交流长廊。此外，设计时我们对整个交通流线进行了深刻的思考，区内车行依托原有控规规划道路在周边形成的环道，通过周边式环状道路和步行轴有效地将汽车交通和人行交通分离，道路尽量形成通而不畅、穿而不透的效果来降低车速，减少噪音，为本区提供安全、宁静、舒适的办公环境。大片绿地景观的运用，营造出一个绿意盎然的生态休闲区；双塔的办公景观是整个区域的制高点，也标示了西南区域形象新高度，世界知名万豪酒店的入驻无疑也是项目的一个亮点，上下双层空间交相辉映，品牌效应的集聚将引爆整个区域商圈。未来，中天金融城商业综合体将成为贵阳城市生活的重要角色，对城市发展产生前所未有的影响。

1 bird's-eye view drawing
2 foot bridge connecting two green spaces
3 drawing of commercial street

1 鸟瞰手绘效果图
2 连接两大型绿地的造型人行天桥
3 商业内街手绘效果图

1	2	5
3	4	6

1 front view of Bank of Communications headquarter
2 surrounding view of the office buildings
3 water features at the main entrance
4 fountain at the entrance of CITIC headquarters tower
5 bird's-eye view in the night
6 night view of the courtyard between headquarters buildings

1 交通银行总部大楼前景观
2 银行办公大楼周边景观
3 主入口特色水景
4 涌泉形式入口设计增强中信银行大厦的活跃氛围
5 夜景鸟瞰图
6 银行总部大楼间中庭景观夜景效果

Zhongtian International Finance Center, Guiyang

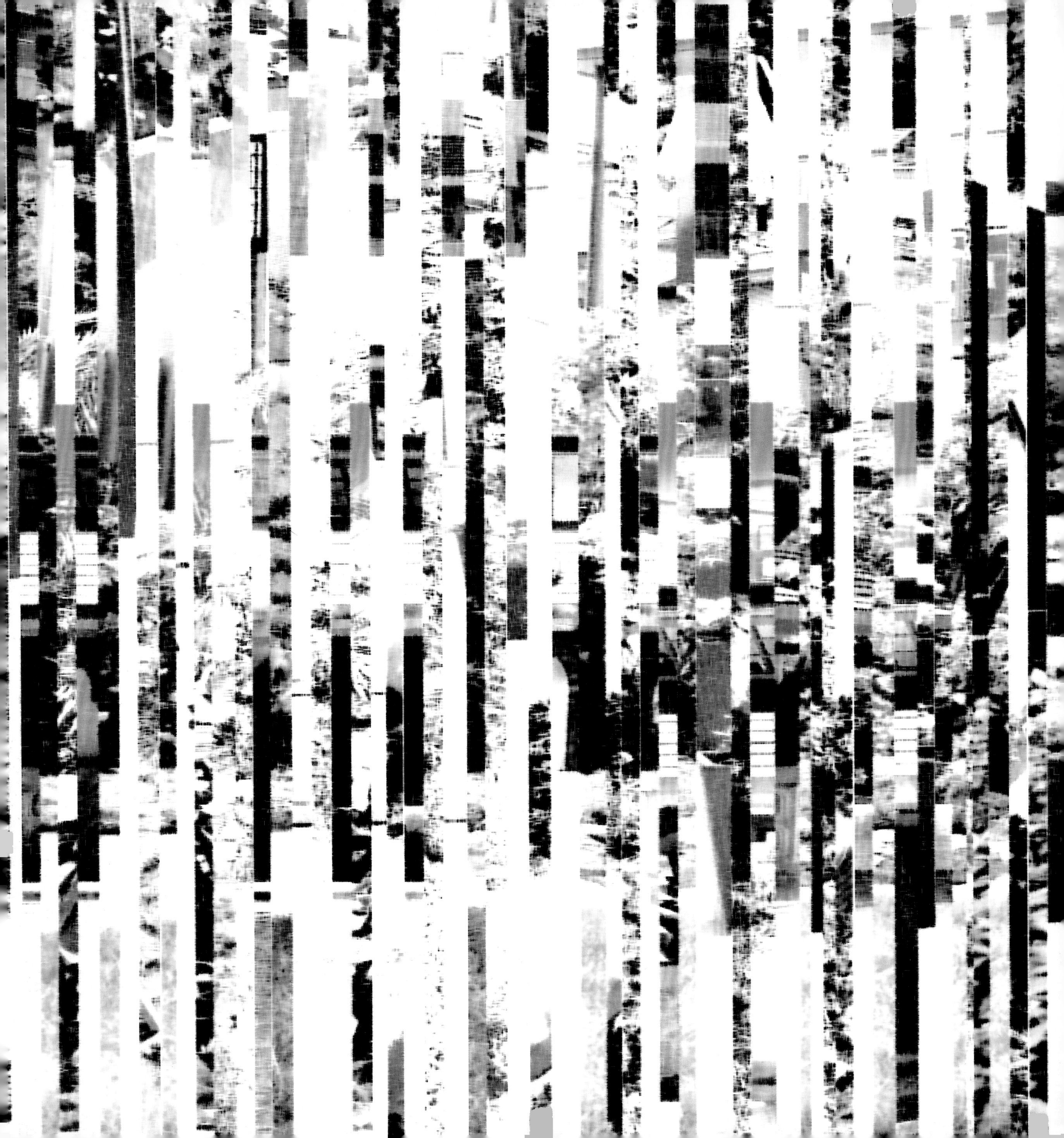

CULTURAL TOURISM +

文化旅游+

Congdu International Conference Center, Guangzhou
从都国际会议中心，广州
——南中国最奢华的高尔夫温泉度假“行宫”

Developer: Qiaoxin Group
Project Type: Holiday Resort Hotel
Project Area: About 200,000 m^2
Design Content: Landscape Detailed Design, Softscape Design & Production Design
Design Period: 2009
Coorperative Unit: BCI

委托单位：侨鑫集团
项目类型：旅游度假酒店
项目面积：约 200 000 m^2
设计内容：景观深化设计、绿化设计及施工图设计
设计时间：2009 年
合作单位：BCI

1 | 2

1 overlooking the buildings of Congdu International Conference Center which nestle among mountains and forest like royal palace
2 site plan

1 远眺从都国际会议中心建筑群，如皇家宫殿般掩映在山林湖泊中
2 总体规划平面图

1 | 4
2 | 3 | 5

1 water features of the conference center
2-3 corner view of the water features
4 panoramic night view of the conference center
5 public swimming pool

1 会议中心水景
2–3 会议中心水景局部
4 会议中心夜景全景
5 公共泳池

Located in Conghua, the backyard of Guangzhou, and nestled on the bank of the beautiful and ecological Liuxi River, Congdu International Conference Center is urban complex integrating international finance conference center, platinum five-star spa hotel, private museum and spa center, etc.

Congdu project has integrated three elements: classical Chinese architectural design, distinguished private space and abundant natural hot springs. It is like a palace sheltered among mountains, springs, forests and lakes solemnly and dignifiedly, creating an elegant shrine atmosphere and inspiring visitors' imagination.

1 | 2 | 3
4

1 elevation of viewing pavilion
2 night view of viewing pavilion
3 magnificent entrance of the villa area
4 section of the landscape at the villa entrance

1 观景亭立面图
2 观景亭夜景
3 恢宏大气的别墅区入口
4 别墅区入口景观剖面图

Congdu International Conference Center, Guangzhou

1	2	4
3		5

1-2 night view of the landscape wall by the swimming pool
3 elevation of the landscape wall
4 elevation of the swimming pool
5 night view of the swimming pool with tall light poles standing like guards

1-2 灯光掩映下的泳池区景墙夜景
3 景墙立面图
4 泳池剖面图
5 泳池夜景——两旁矗立着高大的喷火灯柱，像队列的卫兵

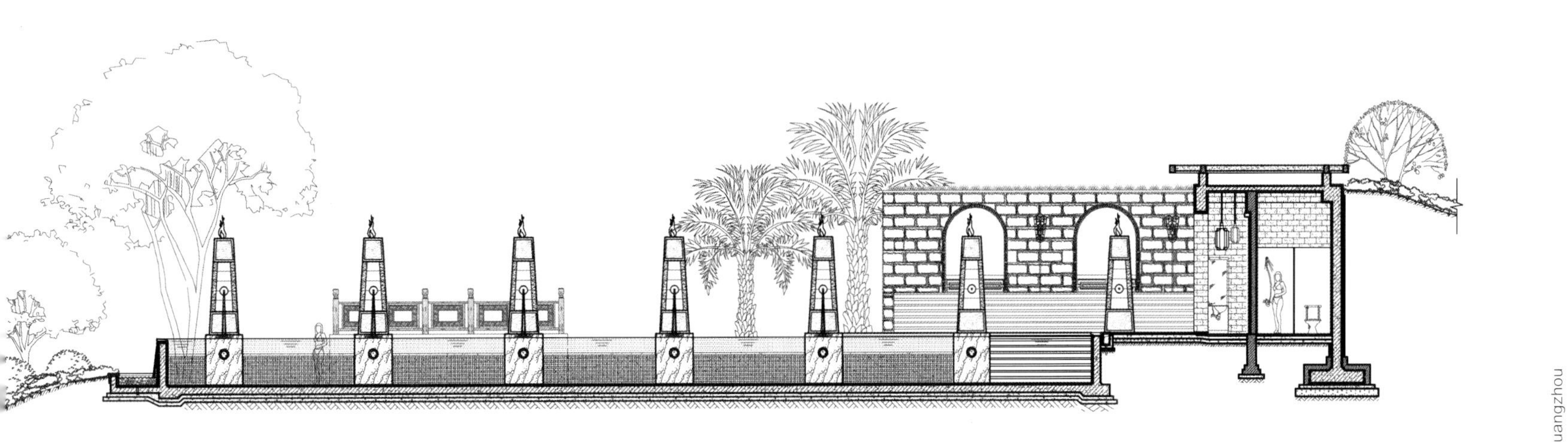

本项目地处广州后花园——从化，坐落于自然风光优美宜人的流溪河畔，项目是拥有金融国际会议中心、白金五星级温泉酒店、私人博物馆及SPA水疗中心等顶尖配套的大型综合体。

从都糅合了三大要素：古典中式设计的建筑、尊贵的私密环境和丰富的自然温泉资源。它宛如一座行宫，气象庄严、雍容华贵地掩映在群山、泉水、森林、湖泊中，完美地与自然融合在一起，营造优雅神宫氛围，让人浮想联翩。

1 | 2
3 | 4 | 5

1 hallway of the courtyard-style suite
2-3 detail drawings of the lights
4-5 landscape items of new Chinese style

1 院落式套房入口门厅
2-3 特色灯具详图
4-5 新中式风格小品

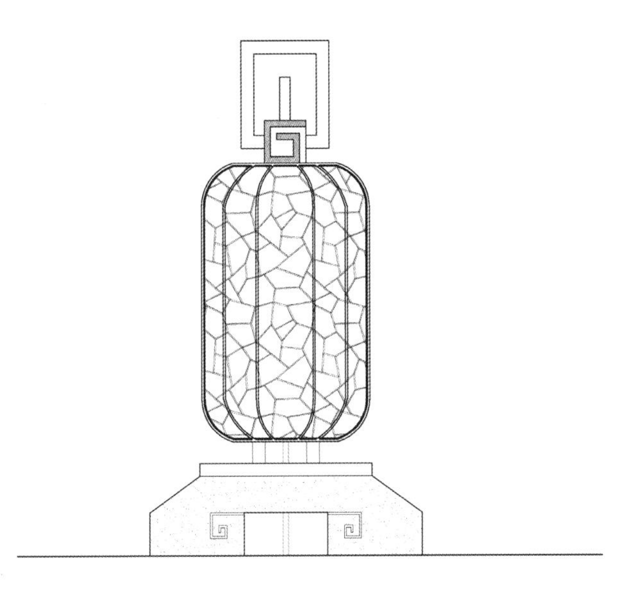

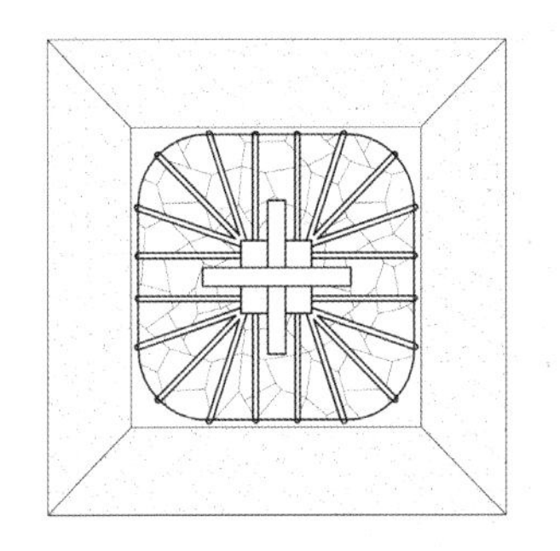

1
2
3

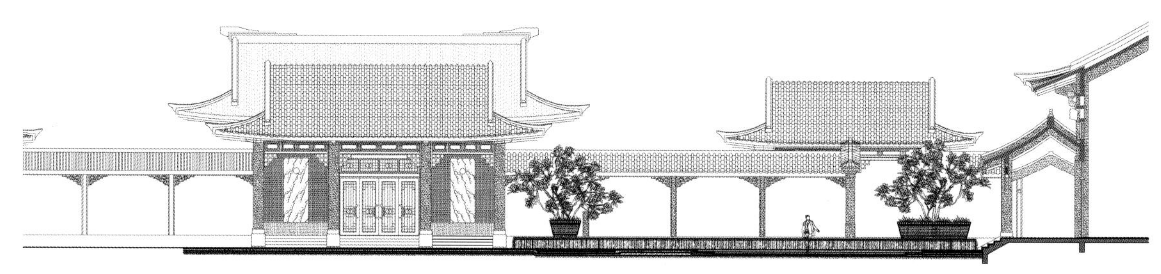

1 classical Chinese-style buildings covered by blue tiles
2-3 elevation of the water feature in the courtyard of the recreational area

1 蓝色瓦面的古典中式建筑，围合出一个尊贵的空间
2–3 会议娱乐区庭院水景立面图

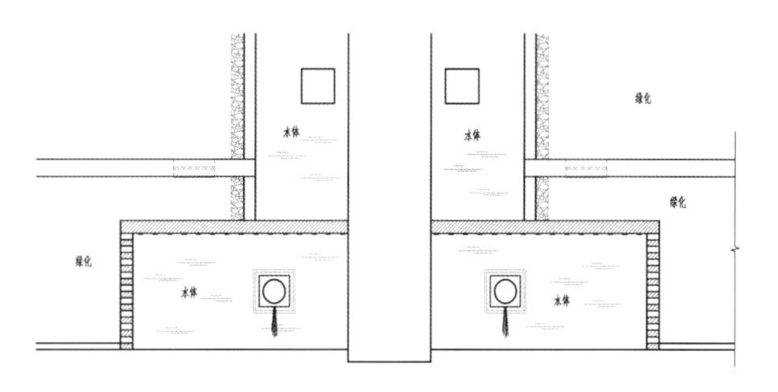

4
5
6

4 floor plan of the waterscape at the entrance of the VIP restaurant
5 floor plan of the waterscape at the entrance of the VIP restaurant
6 landscape at the entrance of the VIP restaurant

4 贵宾餐厅入口水景平面图
5 贵宾餐厅入口水景立面图
6 贵宾餐厅入口景观

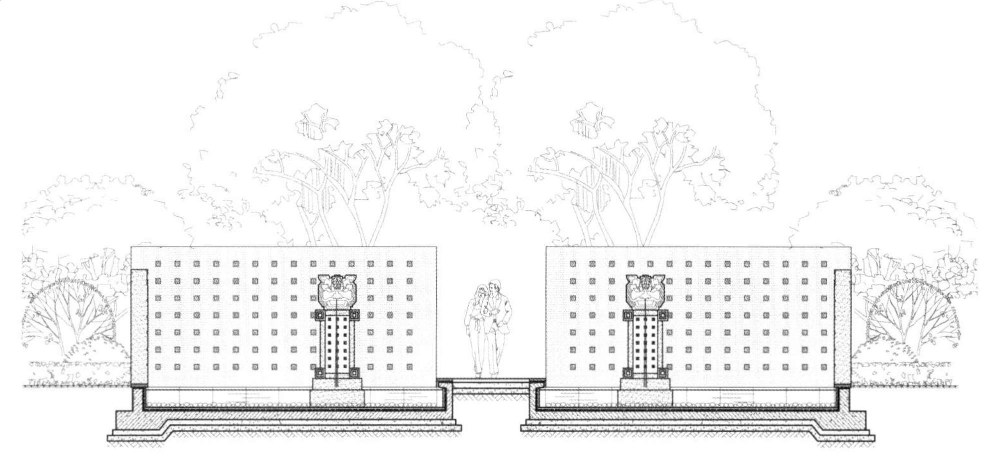

1	2	
	3	4

1 magnificent buildings and multi-level water/planting pools create a comfortable, unpretentious and elegant landscape environment.
2 spacious courtyard space
3-4 swimming pool

1 简洁雄伟的建筑，层次丰富的水池和种植池，营造舒适优雅、低调奢华的景观氛围
2 宽敞的庭院空间
3-4 泳池景观

Flower Town, Guangzhou
花山小镇，广州
——文化寻根，花山艺境

Developer: Guangzhou Kema Real Estate Co., Ltd.
Project Type: Business & Holiday Resort
Project Area: 335,307 m²
Design Content: Overall Plan, Landscape Conceptual Design & Production Design
Design Period: 2014

委托单位：广州科玛置业有限公司
项目类型：商业 & 旅游度假区
项目面积：335 307 m²
设计内容：项目总体规划、景观方案及施工图设计
设计时间：2014 年

Located in Huadu District of Guangzhou City, at Huashan Exit of the Airport Expressway, Flower Town is close to Guangzhou Baiyun International Airport and Huadu downtown, being a cultural and art village characterized by watchtower culture.

At that time, there was a chance to developing tourism villages. Based on the local Lingnan culture, overseas Chinese's hometown culture and watchtower culture, our design team has adjusted the overall layout of Flower Town and built a high-end innovative tourism village which integrates culture, innovation, entertainment, healthcare, resort, business, conference and exhibition.

The landscape is dominated by Lingnan style, showing the charms of diverse culture with details and creating an art village together with the watchtower-style buildings.

项目位于广州市花都区机场高速花山出口，紧邻广州白云国际机场和花都中心城区，是一个以碉楼文化为特色的文化艺术村落项目。

承载着新型城镇化及美丽乡村的发展契机，迎合全国文化艺术乡村旅游发展的热潮，在方案设计中，设计团队对花山小镇的整体布局进行了重新梳理，以岭南文化、侨乡文化、碉楼文化为基础，点线面结合，全方位规划，活化村落，将艺术渗透到场地中，形成三带十区，打造集文化、创意、娱乐、养生、度假、商务、会展为一体的高端创意村落综合体。

景观上以岭南风格为主导，结合碉楼建筑样式，在细节上体现各种文化在场地沉淀下来的魅力，打造多文化汇聚的艺术文化村落，成为极具岭南风情的村落景观。

1 site Plan
2 logo surrounded by flowers and trees
3 watchtower-style buildings hidden in green trees
4 moss and grass on the old brick walls decorate the alley

1 总平面图
2 红花绿树围绕的 logo 标识
3 绿树掩映中的碉楼群
4 古老砖墙上的青苔和小草，给小巷抹上淡淡的绿意

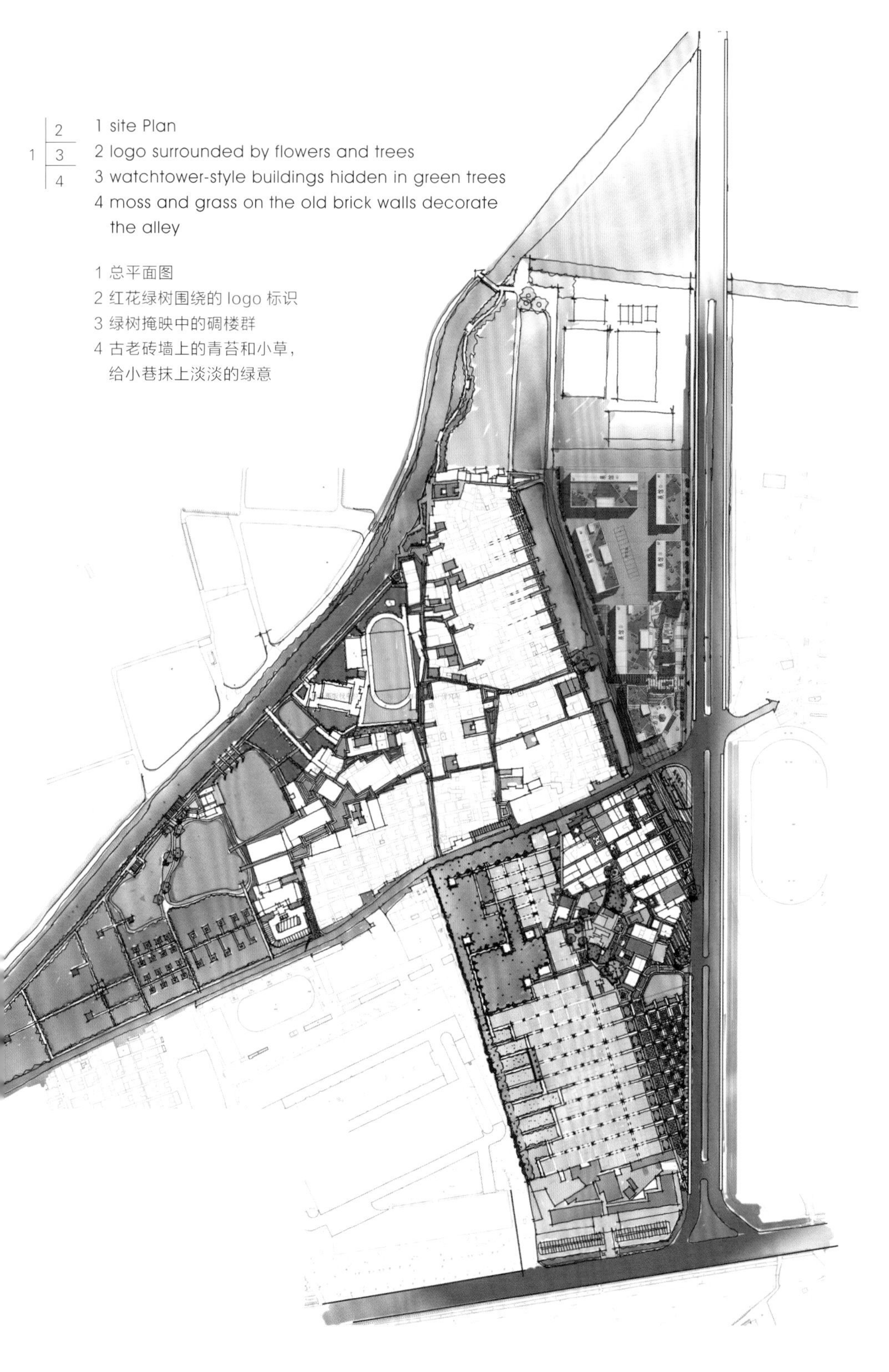

1 | 2
3

1-2 the town is full of life, energy and enthusiasm after renovation
3 colorful and harmonious combination

1-2 改造后的村落焕发新的生机
3 斑斓的搭配耐人寻味却毫不违和

Flower Town, Guangzhou

4
5

4 romantic and nostalgic atmosphere created by soft light
5 lights and mottled footpath

4 柔和的灯光平添浪漫的怀旧氛围
5 斑驳小路、点点灯火透露着浓浓的生活气息

1	2	5
3	4	6

1-3 beautiful town in the night
4-6 new appearance of the small town

1-3 夜幕下的村落别有一番风情
4-6 活化后的小镇旧貌换新颜

Wanda Cultural Tourism City, Guangzhou
万达文化旅游城，广州
—— 水墨生花，晕染盛世

Developer: Wanda Group
Project Type: Commercial and Tourism Complex
Project Area: 25,000 m^2
Design Content: Landscape Conceptual Design & Production Design
Design Period: 2013

委托单位：万达集团
项目类型：商住旅游综合体
景观面积：25 000 m^2
设计内容：景观方案及施工图设计
设计时间：2013 年

Located in Huadu District which is the northern sub-center of Guangzhou City, Wanda Cultural Tourism City includes two parts: the cultural tourism area and the new city. It is envisioned to be an eco residential area and public activity center integrating business office, tourism culture, sports and leisure, commerce and shopping, high-end residences, etc.

The architectural design is inspired by the red kapok - the city flower of Guangzhou. The building blocks are the petals and the podium is the torus. The whole development thus looks like a flower growing from the ground, standing in harmony with the surroundings. The landscape ideas come from the local art - Lingnan painting. Unique design skills are used in creating a modern landscape system which is perfectly combined with the architectures. The whole development looks like a gorgeous Chinese painting with beautiful blooming kapoks. So far Guangzhou Wanda Cultural Tourism City has been the representative of Wanda Cities.

万达文化旅游城位于广州市北部副中心——花都区，项目包括文化旅游区和旅游新城两部分，定位为集商务办公、旅游文化、体育休闲、商业购物和高端居住等多功能为一体的宜居生态居住区和城市公共活动中心的现代服务平台。

本项目建筑形态以代表花城特色的红色木棉花为设计母体进行演变，花瓣为主体，花托为裙楼，整个建筑看起来就像一朵热烈绽放的既坚毅又刚强的木棉花，与周围环境融为一体。景观设计概念源于独特的广州本土元素——岭南画派，汲取其独特的技法运用于现代景观设计中，以花（建筑）为核心，运用撞水、撞粉的写意手法，利用向四周晕染开来的曲线和渗透形式，分隔平面布局，划分功能空间，使整个景观与建筑浑然天成，宛如一幅充满文化韵味且现代、绚丽的水墨画。因此，广州万达文化旅游城成为迄今具有代表性的万达城项目。

1 bird's-eye view of the main entrance of the square
2 kapok flower-shaped architecture at the main entrance of the square

1 广场主入口鸟瞰图
2 广场主入口木棉花建筑

Poly Silver Beach, Yangjiang
保利银滩，阳江
—— 保利地产集团首个滨海旅游地产项目

Developer: Poly Real Estate Group
Project Type: Holiday Resort
Project Area: 1,834,000 m^2
Design Content: A3/GF/Q/H/J/K/L/M District Landscape Conceptual Design & Production Design
Design Period: 2010 as yet

Covering a total land area of 3,000 mu (2,000,000m^2), Silver Beach is the first costal resort project of Poly located in the west of the "Ten Miles Silver Beach", on the south bank of Hailing Island. It leans against Zhuyanding mountain and faces to the vast sea, enjoying incomparable views and resources. The Silver Beach is built to be a top eco destination and high-end recreational community which integrates coastal tourism, sports and recreation as well as healthcare, for tourists and guests from Guangdong, Pearl River Delta and other regions of China.

Inspired by Southeast Asian landscape style, the project has integrated the local culture with maritime elements, and introduced Bali style into the residential area of South China by using typical landscape architectures, landscape items and plants. In this way, residents can experience comfortable Bali lifestyle without going abroad.

The landscape design is based on the overall planning of the buildings. Three landscape axes runs across the villa area to well connect the villas with mountain and sea.

With great respect to nature, the designers have created some artistic landscape architectures to match the mountain view, sea view and plants. At the same time, they insist on the principle of "ecology first" and take advantage of the natural topography and vegetation, creating a natural, ecological and comfortable environment for living and recreation.

委托单位：保利地产集团
项目类型：旅游度假区
项目面积：1 834 000 m^2
设计内容：A3/GF/Q/H/J/K/L/M 区景观方案及施工图设计
设计时间：2010 年至今

作为保利地产集团首个滨海旅游项目的保利银滩，总占地约 3000 亩（200 万平方米），位于海陵岛南岸“十里银滩”西区，北靠岛内两大高峰之一的竹眼顶，南面广阔浩瀚海景，拥有靠山面海的绝佳空间资源。项目旨在塑造面向广东以及珠江三角洲、乃至全国的集海滨度假、运动休闲、康体养生为一体的顶极生态型滨海国际旅游度假目的地和高档休闲社区。

设计通过借鉴东南亚风情的园林设计手法，将项目地域文化和海洋元素相结合，运用景观构筑物、景观小品、特色植物种植等元素营造园林景观，把巴厘岛风情融入到南中国的居住区，让住户感受到度假圣地浓郁的生活氛围，全天候享受具有巴厘岛风情的舒适生活。

景观设计从总体建筑规划布局出发，景观因地制宜，别墅区通过三条景观轴划分出的景观格局，更好地将山体，别墅，海景连成一片，形成视线通廊，三者交相呼应，和谐共生。

设计师以师法自然的设计手法，将山、水、植物与具有艺术内涵的园建小品巧妙搭配，相得益彰。同时坚持“生态优先”的原则，结合地形变化，以自然山水和丰富的植被为主，营造出怡人的自然生态居住空间。

1 elevation of the water feature at the main entrance
2 site plan

1 主入口水景立面
2 规划平面图

bird's-eye view of the site　规划鸟瞰全景效果图

1 | 3 | 4
2 | 5

1 logo at the crossroads
2 artistic boat-shaped cascade
3 luxurious water feature at the entrance
4 details of the water feature and landscape lamps
5 entrance of the sales center

1 路口 logo 标识景观
2 船形跌水充满艺术气息
3 入口水景尽显奢华贵气
4 水景及景观灯具细节
5 远观售楼部入口

国之重器
保国利民

1	2	4
3		5

1 lawn in the backyard
2 innovative sculptures on the lawn
3 dynamic dolphin sculpture in the fountain pool
4 elegant corridor
5 details of landscape lighting

1 后场草坪
2 后场草坪雕塑造型别致
3 喷泉中的海豚雕塑充满动感
4 风雨连廊造型轻巧
5 夜景灯光细节

1			
2	3	4	5

1 waterscape wall at the sub entrance
2 garden lamps
3 textured stone wall in the backyard
4 landscape architectures in the backyard
5 symmetrical fountains in the club

1 次入口水景景墙
2 园灯融入波浪意象
3 后场景观石墙肌理分明
4 后场小品融入场所氛围
5 会所对称涌泉水景

1	4
2 | 3	5

1 commercial street
2 graceful plants in the square
3 green island in the commercial street
4 sign design at the entrance
5 diversified pavement in the square

1 商业街景观
2 广场植物姿态柔美洒脱
3 商业内街个性绿岛
4 入口景观标识
5 广场铺装形式多样

Poly & Shunfeng Bunlos Secret Land, Yangjiang
保利顺峰 · 北洛秘境，阳江
—— 逃离凡尘 恋上桃花源

Developer: Poly Real Estate Group & Shunfeng Group
Project Type: Holiday Resort
Project Area: 22,000 m^2
Design Content: Landscape Conceptual Design & Production Design
Design Period: 2015

委托单位：保利地产集团 & 顺峰集团
项目类型：旅游度假区
项目面积：22 000 m^2
设计内容：景观方案及施工图设计
设计时间：2015 年

Bunlos Secret Land is located in Yangjiang City, Guangdong Province, on the beautiful Hailing Island which enjoys a good reputation of "Beidaihe of South China" and "Oriental Hawaii". Being the first phase of Horsetail Island Tourist Resort, it has taken advantage of the rich natural resources, ocean & silk culture, and ideal location, proposing the planning idea of "paradise with mountain and sea for a free life"and aiming to create luxury vacation residences on Hailing Island.

When entering into Bunlos Secret Land, visitors will be enveloped by a shady, coconut tree-lined boulevard which leads people to two landscape walls made of rustic stones. Then the sight will extend forward along the coconut trees and fountains. At the end of the boulevard, a sunshine lawn is set in front of the sales center to present stylish and multilevel spaces. In addition to modern spaces, traditional steps covered by grass and installed with landscape lamps will greatly improve the visual experience and highlight the unique atmosphere of coastal resort. Behind the sales center is an outdoor swimming pool with simple and elegant lines to outline the blue harbor and dialogue with the unique appearance of the sales center. Shadows of coconut trees, transparent glass facade of the sales center and bright lights reflecting in the water will allow visitors to have a unique and pleasant holiday experience.

项目位于享有"南方北戴河"和"东方夏威夷"之美称的广东省阳江市海陵岛。该项目为马尾岛旅游度假区一期，承接海陵岛丰富的自然资源与浓厚的海洋及丝绸文化，结合场地周边山、海、田围合的自然地貌，提出"山海天境，畅意人生"的规划理念及"沃田近山乐水居，海陵奢美度假游"的项目定位。

初到北洛秘境，呈现在来客面前的是"百步椰林"的入口大道，沿着参观流线步移景异，两面原石朴质的景墙映入眼帘，视线收拢，顺着两边椰林与涌泉一直延伸。连接迎宾大道的是阳光草坪，售楼部下潜的外形承接草坪往上延伸，形成前卫的多层次空间结构，结合精致的嵌灯草阶，现代与传统的转换提升视觉体验，精致的绿化组合与细节处理，无一不体现着度假胜地的独有气质。沿着售楼部到达尽头，便是与天海一色的室外泳池，简洁纯粹的线条勾勒出蔚蓝的港湾，与售楼部独特的外形相映，夜晚椰树林的剪影，通透的售楼部玻璃外墙，灯色辉煌投射在漾漾水中，让来客陶醉在这现代与质朴、精致与风情的感官体验之中。

1 creating a clear and blue harbor under the coconut trees
2 beach chairs by the swimming pool
3 simple and elegant lines

1 沿承天与海的湛蓝，打造一片椰林下的清澈的港湾
2 泳池边上别具风情的沙滩椅
3 简约纯粹的线条

1	2	5
3	4	6

1 green activity space extending to the rooftop
2 exquisitely designed steps covered by grass
3 repetition of different lines creates multilevel spaces
4 entrance space of Southeast Asian style
5 buildings reflecting in the water
6 entrance landscapes in sunny days

1 延伸至屋顶的绿化活动空间，升华建筑前卫的外形构造
2 草阶轮廓融入设计语言
3 不同线条的叠加运用，丰富空间层次感
4 东南亚风情的门庭
5 镜面水与建筑相映
6 晴天下的入口景观

1 colorful lights highlight the entrance space at night
2 coconut tree-lined boulevard
3 blue water tells the story of the Peach Garden

1 夜色里的灯彩蔓延为入口增添氛围
2 椰林夹道风情惬意
3 深蓝的水面静静地讲着远离尘嚣的桃花源

1	3
2	4

1 lights of the sales center reflecting in the water
2 white wall and floor look simple and elegant at night
3 exquisite courtyard lamps
4 steps covered by grass and decorated by landscape lights

1 售楼部的灯色倒映在水中
2 白色的外墙与地面在夜色中突显简单极致的美
3 精致的庭院灯
4 镶嵌灯带在草阶上的运用

Baodun Hot Spring Town, Yingde
宝墩湖温泉小镇，英德
——茶香漫小镇，碧湖有泉声

Developer: Guangdong Richwood Group
Project Type: Hotel & Holiday Resort
Project Area: 895,000 m^2
Design Content: Landscape Conceptual Design & Production Design
Design Period: 2013

委托单位：广东丰泰集团
项目类型：酒店及旅游度假区
项目面积：895 000 m^2
设计内容：景观方案及施工图设计
设计时间：2013 年

Located on the west side of Baodun Lake in Wangbu, Yingde, Baodun Hot Spring Town covers an area of 7,000 mu (4,670,000 m^2) and is about 120 km from Guangzhou, boasts convenient traffic and superior location. It is designed to be a resort town integrating seven functions: culture, entertainment, hotel, commerce, residence, education and office.

In the early stage of the project, we developed a hot spring resort which includes the main entrance, the club, the business area and the holiday resort. Inspired by the local "tea culture", we use "tea leaf" as the foremost landscape element. This design philosophy runs through the entire project, incorporating landscape with buildings and culture. Finally, this culture-inspired hot spring town is presented as a vivid landscape painting under our efforts.

The holiday resort is surrounded by lake. In front of the club, sits an infinity swimming pool, where residents can have a great view of the lake and mountain. Special lanes connected with the pool are made to link the entire resort, thus to let every resident reach the swimming pool just by jumping into the lane in front of their villa.

Designers utilize the narrow space between the villa backyard and the swimming lane and shape it into the form of a tea leaf, providing all villas with leisure terrace and lane inlet. Whilst the combination of spa pool and swimming pool not only creates more function spaces but also makes the most of the hot spring resources.

宝墩湖温泉小镇位于英德市望埠镇宝墩湖西侧，坐拥 7000 亩（467 万平方米）原生土地，距广州约 120 公里，交通便利，区位优势明显。项目规划是集文化、娱乐、酒店、商业、城镇混合住宅、教育及办公等七大功能于一体的度假小镇。

项目早期开发的温泉度假村，包含项目主入口、会所区、经营区及度假区。我们以当地的“茶”文化为主题，以“茶叶”为主要的景观表达元素，以此设计理念贯穿于整个园林中，使景观、建筑、文化相互交融，将地域文化特征浓厚的温泉小镇描绘成一幅气韵生动的山水画卷。

度假村坐拥整个生态湖景，会所前设计了大面积的无边际泳池，将整个湖光山色拥入园中。与主体泳池相连的特色泳道，实现了游泳水系在整个度假村的贯通，让每位住户在别墅门口即可跃入泳道，直达风光无限的无边际泳池，游个酣畅淋漓。

设计利用别墅后院与泳道间的狭窄空间，以柔美的茶叶形态来布局该场地，为每一户设计了休闲平台和泳道下水口。而泡池和泳池的结合不但丰富了其使用功能和空间互动，同时也是将温泉资源做到极致利用的体现。

1 buildings and sculptures of swimming fish
2 bird's-eye view of the infinity pool

1 游鱼雕塑映衬下，主体建筑更显张力
2 无边际泳池鸟瞰效果图

1/2 | 3

1-3 lights in peaceful night

1-3 夜色清静，灯火不眠

Baodun Hot Spring Town, Yingde

宝墩湖温泉小镇，英德

1	2	5
3	4	6

1-3 varied animal sculptures
4 exotic garden lamps
5 dynamic landscape at the main entrance
6 great outdoor views

1–3 造型精巧的各类动物雕塑
4 充满异域趣味的园灯
5 动感张扬的主入口景观
6 户外视野开阔大气

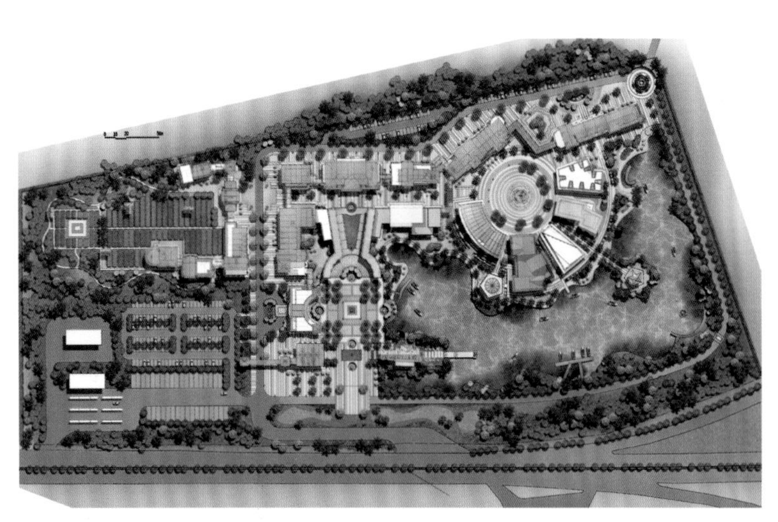

Guangwu Cigar Town, Danzhou
广物・雪茄风情小镇，儋州
——北纬 18° 上的雪茄风情小镇

Developer: Guangdong Guangwu Real Estate Co.,Ltd.
Project Type: Holiday Resort
Project Area: 148,200 m^2
Design Content: Landscape Conceptual Design & Production Design
Design Period: 2013

委托单位：广东广物房地产开发有限公司
项目类型：旅游度假区
项目面积：148 200 m^2
设计内容：景观方案及施工图设计
设计时间：2013 年

Located in the northwest of Danzhou, Hainan, Guangwu Cigar Town is adjacent to Danzhou West Expressway, boasting convenient traffic. Themed as cigar culture, designers incorporate cigar feature with Spanish and North European architectural style to achieve the perfect combination of business, ecology and local culture and to create a high-end tourism town integrating entertainment, leisure, tourism, sightseeing, wedding celebration, etc.

It is a cultural collision and a kind of sense of dislocation through time and space. Various buildings in widely different styles can be seen in this site, such as Cigar Culture Exhibition Center and Tourism Service Center in classical Spanish colonial style, motel in Scandinavia style, and five modern high-quality restaurant buildings by the lake deconstructed by different architectural languages. Rich architectural spaces allows free and expressive landscape design.

Designers try to enable visitors to experience the authentic exotic charm through deep analysis on cigar culture. From the structure, color, size and brand of cigar, to cigar celebrity and to cigar production process and environment, they explored in different angles to seek a creative, artistic and inclusive design technique, i.e., the combination of "deconstructing cigar" and "montage", freeze-framing various cigar scenes at the same space in different time, with different territorial and cultural elements.

项目位于海南省儋州市西北，毗邻儋州市西线高速，交通便利。设计以雪茄文化为主题，以西班牙及北欧风情建筑为基调，融入雪茄特色情调，实现商业、生态、地域文化的完美契合，打造一个集娱乐、休闲、观光、度假、婚庆于一体的高端风情旅游小镇。

一场文化的碰撞和一种时空穿越的错位感：场地内融合了多种风格迥异的建筑，从西班牙古典殖民地风格的雪茄文化展示中心及旅游服务中心，到北欧斯堪的纳维亚风格的汽车旅馆，再到沿湖布置，利用不同建筑语言解构出的五栋现代设计风格高级餐馆。丰富多样的建筑空间，给景观设计带来了富于表现的发挥空间。

通过对雪茄文化的深度解析，力求让游客感受到原汁原味的异国风情。从雪茄的结构、色泽、尺寸、品牌到雪茄名人、雪茄生产过程及环境，我们进行了不同角度的探索，以寻求一种创意性、艺术性和包容性兼容的设计手法，即“解构雪茄”与“蒙太奇”组合运用，利用不同时间、不同地域及不同文化元素，在同一个空间定格出一个个极具雪茄风情的情景。

1
2

1 waterfront commercial street
2 paved square in the business area

1 滨水风情商业街
2 商务区特色铺装广场

Poly Haitang Bay, Sanya
保利 · 海棠湾，三亚
——一个时光融化的地方

Developer: Poly Real Estate Group
Project Type: Business & Holiday Resort
Project Area: 112,041 m²
Design Content: Landscape Conceptual Design & Production Design
Design Period: 2014

委托单位：保利地产集团
项目类型：商业 & 旅游度假区
项目面积：112 041 m²
设计内容：景观方案及施工图设计
设计时间：2014 年

The project is located on Plot C6 of Haitang Bay, Sanya, about 800 meters to the Hainan Branch of PLA 301 Hospital in the northwest, about 2,000 meters to the sea in the east, and close to BT resettlement area on the north. Surrounded by bus station, supermarket, petrol station, etc., it enjoys great convenience. Together with the conference & exhibition city and high-rises nearby, it will become an integrated community with complete supporting facilities.

Characteristics of Housing Products: the low-rise apartment buildings are designed in Southeast Asian style with elegant shape and graceful color. Every villa is designed with private garden and swimming pool on the first floor. Low-density buildings and population ensure a safe and quiet environment. Villas are organized in courtyard layout and divided into groups by roads and footpaths. Every group has its unique sign at the entrance, and the owners can enjoy great privacy at home.

Characteristics of Management: the villas are managed and served by the management company when the owners leave home.

The landscape is designed to follow the style of the architectures. It has created a series of introverted, quiet and private courtyards of modern Southeast Asian style.

The Way to Home: community entrance – 65m two-way road – 55m roadway – 4m road way – 2m footpath. The way to home becomes narrow gradually, leading people back to home and enabling people to get close to nature.

Courtyard Design: every villa group is arranged in courtyard style which enables people to enjoy private swimming pool, quiet courtyard spaces and great privacy.

Details: stone sculptures of Southeast Asian style are simple and natural; tropical species like palms and plumeria are widely planted.

本项目位于三亚市海棠区 C6 地块，其西北距离解放军 301 医院海南分院约 800 米，东面距海边约 2000 米，北面紧邻林旺南 BT 安置区。周围公车站、超市、加油站一应俱全，生活便利。周边建立起来的会展城、远期高层的建设，都将拉动片区的人流量和更多的配套建筑产生，形成一个更完善的社区。

项目产品的特点：具有东南亚气息的公寓，通透轻盈的造型和淡雅的颜色，犹如清新的海风，打造一个精致、舒适的度假社区。公寓：低层休闲公寓，一层配有私家花园及私家泳池，建筑密度及人口密度低。打造一个安静、轻松的度假公寓别墅：便捷流畅的交通将别墅群分成多个组团，每个组团以合院形式组织，各有特色的入口标志区分每个组团，组团内是独门独院的私家别墅，让业主充分享受私密的休闲空间。

管理特点：别墅为产权式度假别墅，管家式管理服务。在业主不在的时候，物业负责打扫庭院，打理一切事物；业主回来的时候，只要提前告知，便会准备好一切让业主宾至如归。

景观设计总体上与建筑风格相呼应，营造内向、宁静、私享的现代东南亚风格院落。

归家序列感：从全区出入口——65m 双向车行道——55m 车行道——4m 车行道——2m 人行步道，通过道路路宽的收窄，让人从城市过渡到家里，增加环境与人的亲切感，归家的感觉。

院落设计：每一个别墅组团都像一个院落，是内聚而不向外开放的，归家之后有私家的泳池、庭院。景观设计延续建筑，注重私密的、宁静不被打扰的空间。

细节：石材雕塑小品的选择是自然贴近原始石材古朴、自然的感觉，有浓郁东南亚风格。植物选择棕榈类、鸡蛋花等具代表性的热带树种。

green vegetation and entrance in the light　层叠绿植和灯光映照下的入口，自有一番静谧的气息

1 waterscape in the night
2-3 villa garden
4 night view of the waterscape

1 夜色中的水景
2-3 别墅花园景观
4 远观水景夜景

Wanda Cultural Tourism City, Wuxi
万达文化旅游城，无锡
——尘俗的忘却

Developer: Wanda Group
Project Type: Commercial and Tourism Complex
Project Area: 29,500 m²
Design Content: Landscape Conceptual Design & Production Design
Design Period: 2014

委托单位：万达集团
项目类型：商住旅游综合体
项目面积：29 500 m²
设计内容：景观方案及施工图设计
设计时间：2014 年

Wuxi, a cultural city in Jiangsu Province with a long history, is place far away from the hustle and bustle of the world. It tells us a legend of Chinese residence — the exhibition area of Wuxi Wanda. Inspired by traditional Chinese gardens and employing new technologies, it has created a new Chinese-style courtyard for modern people. The details of the landscape will showcase the quality of a development. The exhibition area of Wuxi Wanda has well employed traditional landscape skills and perfectly interpreted traditional Chinese culture.

Wuxi Wanda Cultural Tourism City is located closely to the 568-hectare Changguangxi National Wetland Park, enjoying unparallel natural landscape resources. As the focus of the Wanda City, the exhibition area shows respect to the local culture with traditional Chinese-style architectures. Traditional tea ceremony culture is combined with wetland landscape to create a tranquil and elegant environment where people can experience the life of seclusion.

融古今依依风华，在江苏也只有有丰厚地脉文脉滋养的无锡可以承载，在无锡，无需祈望和回眸，掩蔽在太湖的这片土地早已将往事一笔带过，留给我们的是传奇土地的中式大宅。这个传奇就是无锡万达展示区，它采取新的视角，继承中国传统古典园林理念，用最新科技所打造出的适宜当代人居住的新中式院落，在传承与创新之间，承袭盛世国韵。一个楼盘园林景观的细节呈现，总是最能彰显品质。无锡万达展示区继承了传统文化中“移步异景”的造园理念，精雕细琢每处景观，独创出东方禅境园林，成就中国园林经典，诠释代表中国的东方文化之神韵。

无锡万达文旅城紧邻 568 公顷的长广溪国家湿地公园，有水乡泽国，烟雨江南之天然景观优势。展示区作为整个万达城的点睛之笔，用建筑承袭地域文化，烙印在紫砂壶上的千年印记将再次成为这座城市的新名片。茶庭秘境，溪阔水长。我们追随着这份清静、恬澹的东方哲学气质，将茶庭这一源自于茶道文化的独特园林形式与湿地景观融会贯通，以闲寂、优雅的景观氛围，把人们带入一种充满隐逸文化和养生之道的意境之中。

1 | 2
3

1 axial area of the courtyard
2 fragrance and shadows
3 front square (lotus square)

1 石庭小院中轴区实景
2 暗香疏影实景
3 暗香疏影前广场（莲花广场）实景

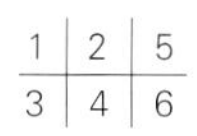

1 luxuriant grass and trees along stone path
2 graceful bamboos at the entrance
3 corner of courtyard
4 garden path
5 pavilion
6 atrium

1 青石路边，草木茂盛
2 门前一丛雅致的竹
3 庭院小角
4 园路小径，纵深感极强
5 荷风四面亭
6 玉堂春中庭

RESIDENTIAL AREA
现代精品居所

Poly Tangyue, Nanjing
保利 · 堂悦，南京
——秦淮厅堂，繁华执掌

Developer: Poly Real Estate Group
Project Type: Integrated Residential Area
Project Area: 76,722 m^2
Design Content: Landscape Conceptual Design & Production Design
Design Period: 2014

The project has won 2014 Meiju Award "The Most Beautiful Residential Landscape"

委托单位：保利地产集团
项目类型：综合居住区
项目面积：76 722 m^2
设计内容：景观方案及施工图设计
设计时间：2014 年

该项目荣获 2014 年美居奖——"中国最美人居景观"

All things past without any trace,
The Residence's owner changed again and again.
When travelers come back home, they will find that
Only the water of Yangtze River remains green.
——"Qinhuai Song" by Emperor Qianlong of the Qing Dynasty

Occupying a total area of 76,722 m^2, Poly Tangyue is located on Plot G48, in the heart of Nanjing South New City. It is planned to be a new center integrating commerce, offices and residences. Designed with classical Chinese elements and modern western techniques, it has created an international residential community in Nanjing City and a new landmark in this region as well.

Poly Tangyue is Poly's first residential complex in Nanjing. GVL has designed its exhibition space as a "living room" to welcome all visitors. Inspired by modern simple style and Chinese Art Deco style, it has employed elegant lines, dynamic experience spaces, hard landscapes, exquisite details and Chinese-style sculptures to enable visitors to experience the charm of traditional Chinese courtyard.

六朝往事难寻迹，王谢燕飞谁氏宅。

风流将令倦游归，唯见长江依旧碧。

——清 · 乾隆《秦淮歌》

保利·堂悦项目位于南京秦淮 G48 地块，属于南部新城中心地段，占地 76 722 m^2，开发定位为集商业、办公、居住于一体的新区域中心。全区以中魂西技为设计理念，运用现代的手法提取古典元素，在展示城市新风貌，打造南京国际社区的同时，注入古都蕴含的气韵，以此传承城市历史记忆，树立区域地标。

本次展现的是 GVL 为保利 · 堂悦项目精心酝酿的展示空间，它以"会客厅"的身份接纳八方来客。这也是深耕商业住宅市场的保利地产在南京倾力打造的首个综合体项目。在设计风格上采用现代简约与中式 Art-Deco 相结合，简约的线条、开合变化的体验空间、反复推敲的硬景体量、精心雕琢的细节组合、错落其间极具中式韵味的主题雕塑等共同营造出千年的秦淮意蕴，也娓娓细说着中国院子的幽幽禅意。

the entrance of the sales center in the night　售楼部入口夜景

保利·堂悦，南京

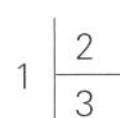

1 the entrance of the sales center in the daytime
2 buildings and waterscape
3 waterscape at the entrance

1 售楼部入口日景
2 建筑与水景一体化设计
3 入口水景烘托活跃气氛

1	2	5
3	4	

1 detailed design can be found everywhere
2 simple and elegant green space
3 skillfully set soft decorations
4 lighting design for better effect
5 exquisitely designed corner

1 无处不在的细节设计
2 简洁大气的绿化空间
3 精心的软装布置点缀
4 利用细节灯光提升整体效果
5 精致端景述说着幽幽的禅意

1 landscape elevation
2 landscape matches the architecture
3 landscape square
4 landscape at the entrance of the sales center

1 震撼的景观立面效果
2 景观设计对建筑效果提升显著
3 动静皆宜的景观广场
4 售楼部入口的精致景观

保利·堂悦

保利·堂悦

1 | 4
2 | 3 | 5

1 modern elegant landscape square
2 planting pool in broken line
3 outdoor fabric decorations
4 exquisite soft decorations and fountains
5 skillful space design

1 现代简洁的景观广场
2 折线种植池缘
3 户外布品融入景观中
4 精致的软装与涌泉相互呼应
5 震撼大气的空间处理手法

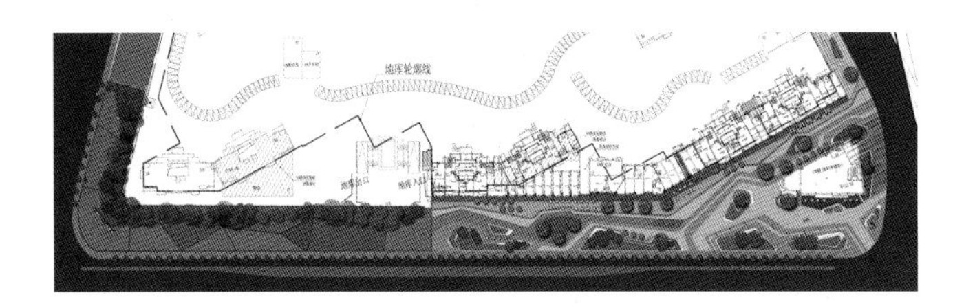

Poly Beautiful River Coast, Nanjing
保利 · 西江月，南京
——U life U like

Developer: Poly Real Estate Group
Project Type: Integrated Residential Area
Project Area: 104,573 m²
Design Content: Landscape Conceptual Design & Production Design
Design Period: 2014

The project has won 2015 Meiju Award "China's Most Beautiful Public Landscape"

委托单位：保利地产集团
项目类型：综合居住区
项目面积：104 573 m²
设计内容：景观方案及施工图设计
设计时间：2014 年

该项目荣获 2015 年美居奖——“中国最美公共景观”

The project is ideally located at the south end of Pukou New Town, Nanjing, Jiangsu, and its demonstration area is close to Nanjing Youth Olympic Games Sports Park. With a total landscape area of 78,150 m², it is planned to be a new regional center integrating commerce and international residential community. The landscape design has followed the modern architectural style, boldly using linear layout, meaningful details and new materials to highlight the uniqueness of this young and stylish international community, and to make it the new residential and cultural landmark of this area.

Based on the architectural planning, the design has followed the architectural style to present modern landscapes, well-organized traffic system, clean and quiet green spaces and diversified pavements.

The space design decides the future public space framework of the community. The design pays attention to people's feeling and creates a people-oriented residential environment. Exquisitely designed function spaces meet the requirements of different people. The spaces belong to six different landscape systems: family activity space, meeting space, parent-child playground, lawn theatre, sports square and jogging track. Every space is carefully designed.

During the process of design and construction, we've considered many pavement materials, and at last, we decided to use the new permeable, decorative and low-cost material --- pervious concrete. We are satisfied with the final effect because of the smooth lines, eco-friendliness and sustainability. Moreover, the carved edges of the tree pool as well as the lighting design also improve the quality of the project.

项目位于江苏南京浦口新城南端，示范区紧邻南京青奥体育公园，地理位置优越，周边交通道路成熟。景观面积 78 150 m²。开发定位为集商业、国际社区于一体的新区域中心。配合建筑的现代风格以及客户群的年轻化，景观延续建筑简约、线性动感的特点，不拘泥于传统设计，在平面布局上注入创新的想法，采用大胆的线性构图，同时融入富有意味的细节，并多处运用新型材料，务求在设计上体现独特性，展现南京现代化城市风貌，营造年轻、时尚，活力的国际社区氛围，使其成为当地居民的居住及人文新地标。

设计伊始，园林设计师结合建筑规划布局，提出设计策略如下：延续建筑手法，营造现代景观；合理组织交通流线，实现人车分流；营造干净纯粹的绿化空间，满足功能需求；注重新型材料运用，强调铺装材质变化。

空间设计是整个社区的骨骼，决定了以后社区的公共空间框架。从人性化角度出发，提供一个切实以人为本的居住区景观。功能空间的精细化提升，根据不同年龄使用人群的需求，考虑其公共、私密的程度，将空间划分为六大景观体系：全家总动员、欢聚天地、亲子乐园、草坪剧场、康体广场、缓跑道。并针对每个空间进行专项设计。

项目从设计到施工过程，在铺装材料的选取上考虑了多种可能性，经多方考量，大胆尝试透水混凝土这一新型材质，透水混凝土不仅具有高透水性、高承载力、装饰效果佳等优点，且成本相较石材低廉，建成的效果也达到了预期，除了将设计的流线感表达得淋漓尽致，更是兼备了生态及可持续性的特点。 此外，树池边缘的精致英文镌刻字体及灯光设计也是本项目的特色。

1
2

1 landscape at the entrance of the sales center
2 reflection pool

1 售楼部入口整体景观
2 镜面水景倒映建筑立面

保利·西江月
千亿钜制 繁华支点
对话河西，南京未来十年经济新增长极
E
B
C
O

1		4
2	3	5

1 bird's-eye view of the recreation area
2 front yard of the exhibition area in the daytime
3 fabric decorations
4 outdoor negotiation area
5 previous concrete square and tree pools in the recreation area

1 休闲活动区鸟瞰
2 展示区前场日景
3 场地软装布品
4 户外洽谈区
5 休闲活动区透水混凝土广场与层级树池

We live in this world when we love it.

1	2	5
3	4	

1 tree pool with hollowed-out rust plate
2 night view of the outdoor negotiation area
3 strip-light tree pool
4 arc-shaped pavement at the entrance of the exhibition area
5 backyard of the sales center in the night

1 树池镶嵌镂空刻字锈蚀板
2 户外休闲洽谈区夜景
3 树池灯带
4 展示区入口处弧线形铺装引导
5 售楼部后场夜景

1 waterscape and tree group at the entrance of the sales
2 night view of the reflection pool
3 modern and elegant landscape at the entrance of the sales center

1 售楼部入口水景树阵与远处精神堡垒组成层次丰富的景观
2 镜面水夜景
3 现代简洁的售楼部入口景观

Poly Beautiful River Coast, Nanjing

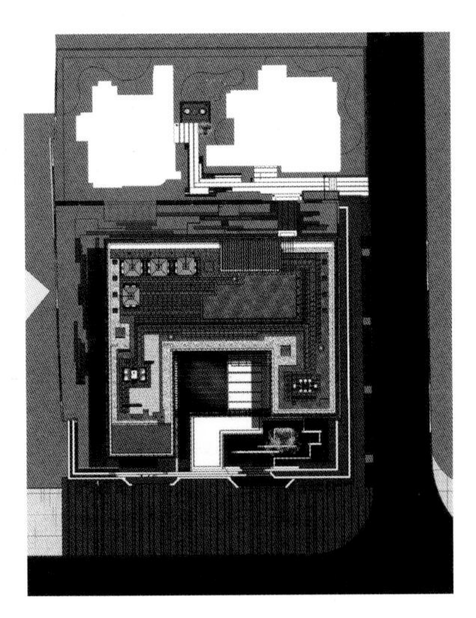

Poly Grand Mansion, Fuzhou
保利 · 天悦，福州
—— 云湖观澜，天悦熙居

Developer: Poly Real Estate Group
Project Type: Integrated Residential Area
Project Area: 30,500 m^2
Design Content: Landscape Conceptual Design & Production Design
Design Period: 2015

委托单位：保利地产集团
项目类型：综合居住区
项目面积：30 500 m^2
设计内容：景观方案及施工图设计
设计时间：2015 年

Located in Fuzhou of Fujian province, surrounded by mountains and waters, this project is just miles away from the West Lake, Zuohai and Pingshan Park, enjoying beautiful landscapes and ideal location. It is also quite near to the government building, Dongjiekou business circle and Wusilu CBD.

The landscape design follows the architectural style to integrate modern style with retro style, enabling the residents to experience exclusive luxury and elegance of hotel style. Natural landscape combines with humanistic feelings to comfort people. Referring to the eight scenes of the West Lake, it has created four landscape areas: water courtyard, stone residence, cloud corridor and moon pavilion.

The front yard of the exhibition area is enclosed by irregular and special-shaped walls, presenting a modern and pure landscape space. Broken lines are employed to make the space more funny, and the cascades make full use of the height difference to enrich the landscape space.

项目位于福建省福州市，坐落于山环水抱之间，距离福州市中心最大的公园——西湖、左海和屏山公园几里之隔，拥有得天独厚的景观、地段及风水优势。省政府、东街口商圈、五四路 CBD 近在咫尺，是坐享城市资源的掌控台。

景观在别具匠心的建筑基础上，精心雕琢，将酒店品质巧妙融入现代都会空间中，形成奢华和内敛雅致的现代触感，现代风格与复古主义相融合，让居者沉浸在专属奢华、高雅所带来的全新感受中，并将自然景观与人文情怀完美结合，达到以景醉人，以境养心。项目结合西湖八景，分别提炼出水庭、石居、云廊、月阁四大景观区块。

项目展示区的前场设计强调空间区域的限定，以不规则、高低起伏的异形墙体营造一个现代、纯粹的景观空间。运用折线的设计手法增加参观的趣味性，层级的叠水不但能令景观更具丰富性，还巧妙地解决了现场的高差问题。

Poly Grand Mansion, Fuzhou

1
2

1 stcps and landscape wall at the entrance
2 scheme effect diagram

1 步级结合入口景墙错落有致
2 方案效果图

landscape matches architecture

景观与建筑相契合

GRAND MANSION
保利·天悦

保利 GRAND MANSION 天悦

保利 天悦
POLY GRAND MANSION

1		4
2	3	5

1 landscape interface in the front yard of sales center
2-3 renderings
4 entrance of the sales center in the night
5 logo wall and waterscape

1 售楼部前场景观界面
2–3 方案效果图
4 售楼部入口夜景
5 logo 标识景墙结合水景

1	2	4
3		5
		6

1-2 rendering of the roof garden
3 flowerpots and lamps in the roof garden
4 lamps of the roof garden under sunset
5 pergola at the entrance of the roof garden
6 enclosed quiet space of the roof garden

1–2 屋顶花园效果图
3 屋顶花园花钵与灯具
4 夕阳下的屋顶花园灯具
5 屋顶花园出入口廊架
6 屋顶花园围合静谧空间

Poly Westriver Imagination, Fuzhou
保利 · 西江林语，福州
——山水文化大宅，出则繁华，入则静谧

Developer: Poly Real Estate Group
Project Type: Large-Integrated Residential Area
Project Area: 141,301.8 m²
Design Content: Landscape Conceptual Design & Production Design
Design Period: 2013

委托单位：保利地产集团
项目类型：大型综合居住区
项目面积：141 301.8 m²
设计内容：景观方案及施工图设计
设计时间：2013 年

As the first luxury mansion development in Fuzhou, Poly Westriver Imagination is located in Huai'an ancient land, to the north of Fujian Agriculture and Forestry University, enjoying the local intelligence and wealthy, and one thousand humanities. Under the city planning of both sides of Fuzhou Minjiang, the surrounding residential commercial facilities of the project become more perfect. It sites very closely to the Hongshan Bridge, 10 minutes directly to the Sam Club, 20 minutes to the East Street Conner, Wanbao business circle and CangShan Wanda, 15 minutes to WuSi Road, Hualin Road and the railway station through the inlets and outlets of Hongtang Town, which holds flourish and nature under control.

The project carries on the habitation spirit of "harmonious nature and lofty life goal", which takes the New York Washington mansion community as the blueprint to refine the luxury residence model scarce in China. It initiates the stereoscopic landscape garden, perfectly combines the building, community and landscape, and extends the charm of the landscape to every corner of the community, which fully explains the landscaping essence of "winding path leading to a secluded spot and harmony between the heaven and human". Whether a stone path, a bunch of flowers, or a garden lamp can keep people far away from the blatant city and regain the touching emotion brought by the quite landscape scenery.

福州首座"清豪宅"——保利·西江林语，坐落于福州淮安古地，福建农林大学北侧，饱享两江灵性润泽，千年人文氤氲。在福州闽江两岸的大都市规划下，项目周边住宅商贸配套愈加完善。咫尺洪山桥，10分钟直达山姆会员店，20分钟连接东街口、万宝商圈和仓山万达，从洪塘出入口接驳三环，15分钟畅达五四路、华林路和火车站。繁华与自然，尽在掌握。

项目传承"和谐自然、宁静致远"的人居精神，以纽约华盛顿清豪宅住区为打造蓝本，淬炼中国稀缺资源豪宅范本。首创立体式山水园林，让建筑、社区和山水完美融合，把山水的魅力，延伸到社区的每一个角落，充分诠释了"曲径通幽，天人合一"的造园精髓，无论是一条石板小路，一簇清雅繁花，还是一盏园林青灯，皆能让人真切感受到远离城市喧嚣，重回山水宁静的久违感动。

Poly Westriver Imagination, Fuzhou

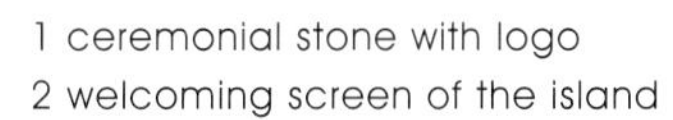

1 ceremonial stone with logo
2 welcoming screen of the island

1 标识 logo 景石增强仪式感
2 岛屿一侧屏风加强迎宾仪式

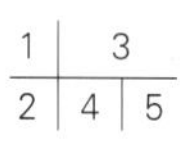

1 swimming pool and cascade
2 steps and tree pools
3 grass steps and vegetation
4 landscape architectures by the swimming pool
5 landscape details

1 泳池结合叠水景观
2 台阶树池井然有序
3 草阶植被烘托空间感
4 泳池两侧风情景观架构
5 景观细部

Poly Westriver Imagination, Fuzhou

1	4
2 | 3	5

1 panoramic view of terrace waterscape
2-3 leisure platform
4-5 details of waterscape

1 主轴台地水景全景
2-3 休闲平台
4-5 水景细部

1 | 2 | 3
--- | 4

1 steps, landscape wall and lamps
2 innovative landscape lamps
3 garden lamps
4 waterscape in the front square of the exhibition area

1 台阶结合景墙、灯具
2 创意景观灯
3 景观庭院灯
4 展示区前广场水景

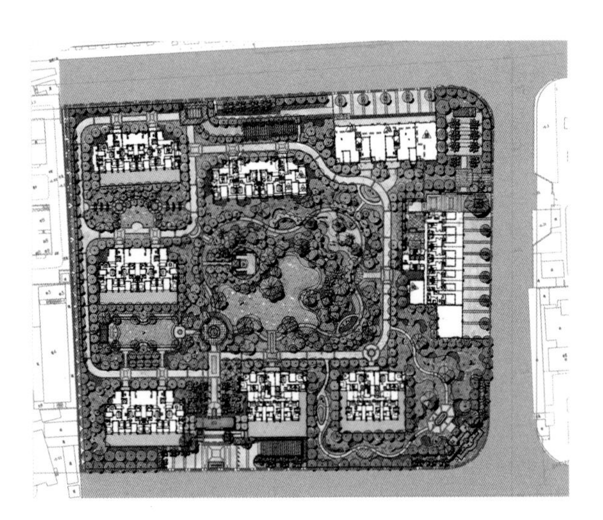

Zhonggeng Premier Luxury River, Fuzhou
中庚·香江万里，福州
——CBD 城市公园华宅

Developer: Fujian Zhonggeng Real Estate Co., Ltd.
Project Type: Integrated Residential Area
Project Area: 38,281 m^2
Design Content: Landscape Conceptual Design & Production Design
Design Period: 2013

委托单位：福建中庚置业有限公司
项目类型：综合居住区
项目面积：38 281 m^2
设计内容：景观方案及施工图设计
设计时间：2013 年

Located in busy Taijiang District of Fuzhou, Fujian, adjacent to Wanda Plaza and International Strait Convention Center, Premier Luxury River enjoys convenient traffic and favorable conditions for a modern high-quality residential community. Buildings are designed in neo-classical Art-Deco style to be modern and dignified. Beige is the dominant color, while brown and dark brown are used to highlight the contours of lines. The landscape design follows the modern architectural style to create romantic eco spaces with forest and lawns. Details such as the pavements, landscape architectures, decorations, landscape walls and materials are exquisitely designed to highlight the quality of the community. Simple and elegant design not only meets the functional requirements but also create different landscape spaces with soft vegetation.

Landscape design of the 6,792 m^2 exhibition area follows the architectural style and extends the architectural spaces. Eco ideas combine with details to present the beauty of exquisiteness and harmony.

The elegant main entrance space of the exhibition area is designed with landscape walls and trees to enhance the ceremonial atmosphere. Natural zigzag garden path leads visitors into the sales center, while the reflection pool and fountain on the other side highlight the commercial atmosphere. Exquisite landscape architecture in the center enriches the simple façade and highlights the elegant design idea.

Fuzhou is a city without scorching summer and freezing winter. There are flowers and green plants in all seasons. The plant design of the exhibition area shows the climatic feature, creates a natural environment and highlights the elegant community culture. Trees are planted in line to enclose the square together with lawn and flowers. Colorful flowers on the lawn look like a colored ribbon, natural and romantic. The meeting area is paved with grass and decorated by beautiful plants to provide a comfortable space for talking.

项目位于福建省福州市台江区，地处繁华区，毗邻顶级城市综合体万达广场及海峡国际会展中心，交通便捷，同时为本案打造现代高品质生活社区提供了优良的条件。建筑为新古典（偏现代）装饰艺术风格为基调，突出现代的尊贵感。色彩以米黄为主，褐色和深棕色强调线条感。景观设计延续建筑现代感及尊贵感，通过密林及草坪营造浪漫的生态空间，通过铺装、构筑、小品、景墙、材料质感等细节方面突显社区品质。景观设计在满足人们日常生活要求功能的基础上，尽量采取简洁大气的设计手法，减少硬质铺装的使用面积，通过植物造景，围和出不同类型的景观空间。

本项目展示区用地面积为 6 792 m^2，景观与建筑风格相呼应，同时对建筑空间进行了延伸，将生态自然的理念与精致的细节相结合，展现一种精致、和谐之美。

1
2 | 3

1 fountains of the sales center
2 landscape wall at the entrance with flowers
3 neoclassical garden lamps match the waterscape

1 售楼中心特色喷泉水景
2 入口景墙前花团锦簇
3 新古典式园灯契合水景

展示区精致大气的主入口空间，采用景墙结合树阵的形式强调仪式感，内部主要以蜿蜒自然的园路强调自然优雅的游览感受，并在一定程度上使售楼部建筑得以凸显，另一面镜面水景结合涌泉的形式烘托售楼部的销售气息，同时以精致的构筑物作为视觉焦点，使简洁的售楼处景观在立面上得以丰富，与整体典雅精致的设计理念相呼应。

展示区绿化景观以体现福州“四季常青、四时开花；夏季无酷暑、冬季无严寒“的宜人居住环境，展现简约大气的社区文化，同时兼顾销售特色，营造清新自然的生活场景为目标。植物搭配上，主要运用景观轴线对称方式，以树阵为主要配置手法，搭配草坪、鲜花，延续轴线景深，在满足树荫广场的功能基础上，着重凸显展示区入口的气氛与尊贵感。展示区中间大草坪区域上设计不同颜色的花卉，犹如彩带般绚丽，在感受开阔大空间的同时，领略大自然的烂漫气息。销售洽谈区通过景观视线延伸到舒缓的地形，适当营造草坪空间，同时利用植物的线条和构图形式优美的林冠线和林缘线，提供舒适、轻松的洽谈环境。

1 | 2 / 3

1-3 quiet leisure & meeting area in the sales center enclosed by pergolas

1-3 售楼中心廊架围合出安静的休闲会客区域

4 | 5
6

4 deep and quiet garden path
5-6 meeting space of the sales center enclosed by pergolas

4 特色园路格外幽深
5-6 舒适自然的开放式草坪

1	3	
2	4	5

1 night view of the pergolas and waterscape in the sales center
2 pergolas of the sales center
3-5 pergolas in the night

1 售楼中心廊架夜景水景
2 售楼中心廊架
3-5 廊架夜色撩人

Poly City, Putian
保利城，莆田
——新贵时尚大都会

Developer: Poly Real Estate Group
Project Type: Large-scale Integrated Residential Area
Project Area: 193,889.75 m^2
Design Content: Landscape Conceptual Design & Production Design
Design Period: 2015

委托单位：保利地产集团
项目类型：大型综合居住区
项目面积：193 889.75 m^2
设计内容：景观方案及施工图设计
设计时间：2015 年

Poly City is located in the center of Hanjiang District of Putian City. Taking advantage of the riverbank location, it is built to be an eco, livable, high-quality and high-end residential community. The overall layout is composed by natural curves and simple geometries. The ceremonial landscape axis and the open central garden interprets cozy lifestyle. Diversified spaces will change according to different seasons. Romantic flowering plants make the modern buildings more dynamic. Details like garden paths and decorations are specially designed for the aristocratic families on the bank of Hanjiang River.

莆田保利城位于莆田市涵江区的核心地带，项目依托原场地自然水岸的资源优势，同时注入现代都市元素，打造一个“生态、宜居、品质、尊贵”的中高端时尚住宅。整体布局以自然的曲线融合简洁的几何构图。主从分明的中轴线奠定清晰的礼仪感，大面积的开放式中央花园演绎惬意的生活气息。空间变化富有层次，充满四季的变化感。现代式建筑在浪漫的多花植物的映衬下更显活力气息。特色园路、园林小品等高品质细节，符合专属定制的望族气质，成就涵江边阳光多彩的尊居家园。

Poly City, Putian

1
2

1 landscape wall with logo at the entrance
2 dynamic fountains in the night

1 入口 logo 景墙大气时尚
2 夜色中的涌泉极富动感

保利城，莆田

1 | 3
2 | 4

1 lighting design at the entrance
2 details of the planting pool
3 bird's-eye view of the exhibition area in the night
4 bird's-eye view of the exhibition area

1 入口灯光设计
2 种植池细节考究
3 展示区局部夜景鸟瞰
4 展示区局部鸟瞰

样板间
Sample Area

1	2	5
3	4	6

1-4 detail design can be found everywhere
5 dynamic fountains echo elegant architectural facade
6 mulli-level landscape

1-4 随处可见的细节设计
5 沉稳大气的建筑立面与动感的涌泉互相映衬
6 富有层次的景观

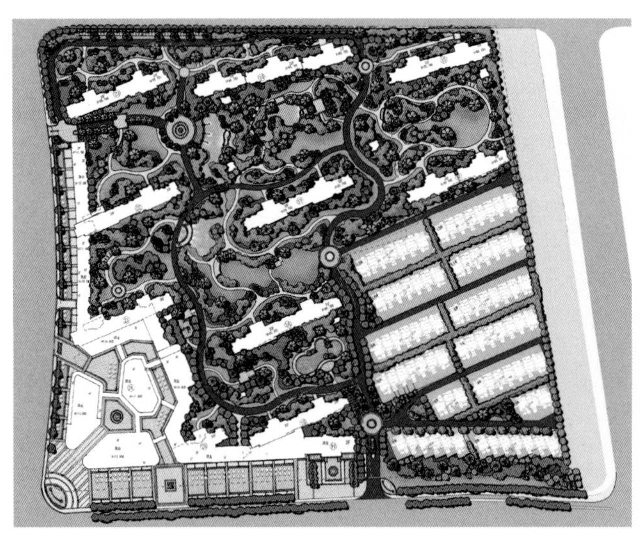

Jinke Lounge Impression, Zhangjiagang

金科·廊桥印象，张家港

——缤纷廊桥，绽放在托斯卡纳的艳阳下

Developer: Jinke Real Estate Group
Project Type: Integrated Residential Area
Project Area: 154,274 m²
Design Content: Landscape Conceptual Design & Production Design
Design Period: 2012

委托单位：金科地产集团
项目类型：综合居住区
项目面积：154 274 m²
设计内容：景观方案及施工图设计
设计时间：2012 年

Ideally located next to the new government building of Jingang Town in Zhangjiagang, Lounge Impression is near to Xiangshan Scenic Area, being the largest and the most luxurious housing development in this area.

Design Concept: bathing in the sunshine of Tuscany is a great moment for everyone. Everyday we will wake up in the morning sunshine of Tuscany and walk on the Tuscan land in the daytime. The air will refresh you and remind you of the happiness of life. It is what life is like in Lounge Impression. The elegant Italian style rebuilds the natural landscapes and presents a green and sunny home with European style landscapes.

The square in front of the sales center is well paved with European style fountains in the center to highlight the sense of dignity.

The romantic lake lawn is full of beautiful flowers; iron tables and chairs, terra cotta flowerpots and natural rocks on the lawn show the charm of Tuscany style.

The dark blue swimming pool connects with the lawn, and the architectures of graceful Art-Deco style hidden among the pine and fir trees. The overall layout of the garden is natural and reasonable, providing different views step by step.

金科·廊桥印象位于张家港金港镇新政府大楼旁，邻近香山风景区，地理位置优越，是目前当地规模最大、规划起点最高的首席高档楼盘。

设计理念：守候着托斯卡纳的艳阳，连时光都变得如此缤纷而畅美。希望每一个黎明，都能在托斯卡纳的艳阳中醒来；希望每一个白天，都能穿梭在托斯卡纳起伏的大地上。这里的空气似乎能穿透你的心扉，这里有你一贯追求的理想生活，这里的一切都是正好，这就是金科·廊桥印象。我们以“精致写意”的意大利情怀为风格立意，通过对自然美景的重塑，结合趣味生动的欧式景观小品层层烘托，力求打造一个“绿意盎然”“优雅阳光”的生态家园。

售楼处前广场以大面积铺装设计为主，中心点缀欧式喷泉景观，带给人气势强烈的尊贵感。

在清新浪漫的草坪湖区，开得正艳的时花带优美如练，点缀其中的铁艺桌椅、赤陶花钵、天然石块等景观小品更是处处体现着浓郁的托斯卡纳风情。

湛蓝的泳池与宽敞的草坪相接，典雅的 Art-Deco 建筑在彰显着托斯卡纳风情的松柏中若隐若现，整个园区布局自然、移步易景，让人流连忘返。

rolling micro topography, romantic lawn lake　起伏的微地形，浪漫的草坪湖

1 | 3
2 | 4

1 deep blue swimming pool with flower patterns on the bottom
2 swimming pool connects with lawn lake through zigzag garden path
3 recreation area with recliners
4 section of swimming pool

1 湛蓝的泳池，池底有雍容典雅的花式图案
2 泳池与草坪湖边缘通过曲线形的园路自然相接
3 躺椅休闲区
4 泳池手绘剖面图

1	3	4	
2		5	6

1 bright red flower belt looks like running stream
2 garden path around the lawn lake
3-6 landscape details of the showflat in villa area

1 鲜艳的红色花带像灵动的溪涧
2 徜徉在草坪湖里的休闲园路
3-6 别墅区样板房细部景观

Everbright Top of the World, Dongguan
光大 · 天骄峰景，东莞
—— 演绎新东方大宅华景

Developer: Guangdong Everbright Enterprise Group
Project Type: High-density Residential Area
Project Area: About 133,300 m²
Design Content: Landscape Conceptual Design & Production Design
Design Period: 2005

委托单位：广东光大企业集团
项目类型：高密度居住区
项目面积：约 133 300 m²
设计内容：景观方案及施工图设计
设计时间：2005 年

The project has won the silver prize of the 5th Huacai Award (landscape design category) in 2015; International Landscape Award issued by BALI in 2013; the first prize of Guangdong Excellent Engineering Survey and Design in 2013; Meiju Award "China's Most Beautiful Residential Landscape" 2013; the first prize of Guangzhou Excellent Engineering Survey and Design in 2012; "2010 China's Low-carbon Apartments"on the 11th China Real Estate Development Meeting; innovation prize of Global Happy Residential Community in 2007; 2007 State-level Green and Health Apartments; 2007 China Top Downtown Residence; 2007 Dongguan's Best Landscape Residence; 2007 Dongguan's Gold Prize Landscape (apartments); etc.

该项目荣获 2015 年第五届"华彩奖"园林和景观设计类银奖；2013 年度获英国国家园景工业协会（BALI）国际园景项目金奖，广东省优秀工程勘察设计奖一等奖，美居奖——"中国最美人居景观"；2012 年度获广州市优秀工程勘察设计评选园林组一等奖；2010 年在第十一届中国房地产发展年会上被评为"2010 中国房地产低碳示范楼盘"；2007 年"全球人居环境幸福社区创新奖"、"国家级绿色健康宜居示范楼盘"、"中国城市中心区首席名宅"、"东莞最佳园林顶级豪宅"、"东莞金牌地产评荐金牌园林（洋房）"等多个奖项。

landscape at the entrance 入口景观

1. Site Condition and Overall Plan

Situated in East Dongcheng District of Dongguan City, with Huangqi Mountains on the south and adjacent to Huying Reservoir and Fengjing Golf Court, the development enjoys superior location and advantaged landscape resources. This long and narrow site occupies an area of 167,046 m^2, on which the total floor area is 508,372 m^2. Building groups are well arranged in ring from east to north to get the best golf views in the south. And villas are superposed in form of highrise to overlook the beautiful landscapes. It takes advantage of the topography to build the semi-underground parking in the low-lying land. The 100,000 m^2 Tianhu Garden and the 40,000 m^2 artificial lake are built on flat land. Great lake views will give strong visual impact. Based on the theme of "mountains, water and nature" and with the innovative design of materials, the garden is designed in natural style to bring people close to nature.

2. Landscape Planning Ideas and Practice

2.1 Landscape Resources Borrowing and Creation

The landscape planning pays attention to the connection between indoor and outdoor spaces, and keep the artificial landscapes in harmony with the surrounding environment (mountains, lake, golf court and native plants). Buildings are arranged in row to face south, which enable people to enjoy beautiful golf views in the southeast and misty mountain views far away. Without any obstacles, one will feel pleasant to overlook the 10 km landscapes. To complement the breathtaking natural landscapes, the designers have designed an ecological waterscape garden which well integrates into the natural environment to present perfect landscape effect.

2.2 Culture Expression and Cultural Atmosphere Creation

There are two ancient banyan trees standing in mountain park on the north of

Scenery of main entrance lanscape
主入口景观实景

光大·天骄峰景，东莞

the site. They are more than 200 years old with thick trunks, stretching canopies and luxuriant foliage. As time passes by, everything changes except two banyan trees. So the designers have kept these two trees and collected memories about them to create a series of meaningful spaces that embody local culture. In these spaces, plants and landscape architectures act as the culture carriers beyond their physical functions.

2.3 Reservation of Native Landscape and Creation of New Landscape

Useful native plants are reserved and play an important role in the landscape design. A lot of maples, frangipani, Alstonia scholaris, Bauhinia blakeana and other native plants are planted to echo the ancient banyan trees. In this way, it not only saves purchasing and maintenance fees but also creates a familiar and friendly environment. In terms of hard landscape, it mainly uses cheap and locally processed stones in ground pavement and landscape architectures.

2.4 Chromatography Design of Landscape

The designers have drawn a chromatography for the landscapes to control the overall landscape effect, and at the same time to use colors to shape unique identity of this development. Gray and green are dominant colors which appear as large-area plants and paved roads. While small spaces are characterized by bright colored flowers and plants, red brown buildings, bronze sculptures or yellow terraces. Dignified and dynamic, it aims to create a harmonious and peaceful environment for this community.

2.5 Landscape Axis

2.5.1 East-West Axis

In the east-west direction, it is a waterscape axis. The artificial lake (total area: 18,000 m^2; biggest width: 70 m; average width: 35 m; shoreline: 600 m) has become the focus of the whole community which is built above the underground parking. The lake forms a clear landscape context together with the surrounding plants and green lands. Around the lake, squares, decks, wooden trestles, streams, fountains and springs are designed to provide diversified waterfront spaces for playing and relaxation.

landscape of zen style at the entrance
入口景观融揉禅宗意境于叠山理水之中，既俊逸秀雅，又不失尊贵大气

2.5.2 South-North Axis

In the south-north direction, it is a natural art axis which starts from the landscape square at the south entrance, extends through the bridge over the lake and connects with the mountain park on the north. With elevation difference between landscape square and the underground parking, it creates an entrance that connects with mountain path, lake views and golf views. Waterfalls and streams flowing among stones, hundred-year-old Podocarpus macrophyllus and royal Pinus thunbergii Parl standing like guards, all combine together to make the entrance elegant and dignified. At the same time, it takes advantage of the topography and plants to build a mountain park back on the north, which provides the residents an ideal place to have fantastic experience.

2.6 Plant Arrangement

The plant arrangement is inspired by the four rules of tea ceremony: harmony, respect, limpidness and quietness. Open forest and lawn on the lakeside provide great waterfront spaces: tall trees divides the lawn into small diversified spaces. In the places where high density is needed, different species are mixedly planted, namely, trees, bushes and vines, evergreen plants and deciduous plants, or rank vegetation and slow-growing vegetation. This kind of arrangement forms multi levels and different phytocenosis. Along the shore, there are mainly weeping willows as well as water plants such as arundo donax var. versicolor and cyperus altrnlifolius to form interesting wetland views.

2.7 Landscape Details

Community enclosure: 800 m long and 4.5 m high embossed enclosure with copper inlaid and hand paintings. It's now applying for the National Patent.

Moon Pagoda: this 10 m high pagoda stands in the southwest of the garden, which is designed in modern and elegant style, overlooking the surroundings.

Copper sculptures: sculptures of different styles are designed to decorate different spaces. Red-crown crane sculptures stand at the entrance; other animal sculptures create an interesting art garden; varied spraying sculptures and music sculptures greatly upgrade the level of the community.

1	3
2	4

1 entrance stone
2 entrance waterside pavilion
3 main-axis landscape bridge
4 elevation of the main entrance

1 入口标志景石
2 入口水景处水榭景观亭
3 主轴景观桥
4 主入口立面图

1 建设地现状和总体布局

项目位于广东东莞东城区东部，南有黄旗山脉，毗邻虎英水库和峰景高尔夫球场，地理位置和景观资源十分优越。建设基地 167 046 m²，总建筑面积 508 372 m²，地形狭长，南北进深小。建筑规划布局最大化地利用地块南面高尔夫景观资源，整体建筑群呈东西横向环抱式排列，使得景观视野开阔不受阻碍。采用高层别墅建筑形式，具有"会当凌绝顶，一览众山小"的俯瞰景观视野优势。在控制成本的前提下发挥最佳的景观效果，利用地势低洼的区域建造半地下车库，而在地面部分则营建小区内 100 000m² 的天湖园林景观和一个面积达 40 000m² 的大型人工湖。园林以"亲山亲水亲自然"的景观主题、自然氛围的写意风格，将材料设计作为创新的主调，让大湖面景观带来视觉冲击和尊贵感，给人们回归自然的感觉。

2 景观规划构思要点及做法

2.1 景观资源的因借与塑造

园林景观规划构思时非常重视内外空间的紧密联系和相互渗透，将湖、山、高尔夫球场、良好的植被等已有的自然景观和人工景观资源充分融入，将建筑按照面向南面的方式一字展开，将东南面的高尔夫球场起伏流畅的草坪地形景观和远处薄氲中的山色纳入到居住区的整体景观中，模糊了区内与区外的界线，远眺 10km 的观景视野亦让人心旷神怡。为呼应区外的壮美景色，营建了一个能媲美大自然的生态水景园林，使景观近大远小的透视效果更贴近视觉尺度，并和外界自然空间完美的融合，形成双园林的景观效果。

2 | 1

1 large area of lake view is the focus of the project
2 In terms of plant arrangement, the spiritual connotation of tea ceremony is used, like "harmony, respect, limpidness and quietness", to enhance the spiritual quality of this property

1 大面积湖景是项目的焦点
2 植物配置借用了茶道"和、敬、清、寂"的精神内涵，注重提升楼盘精神素质

2.2 住区内文化意象表达和意境营造

项目北面的山体公园上屹立着两棵古榕树，躯干粗壮，树冠舒展，枝繁叶茂。这两棵古榕树已站立在山坡上 200 多年了，年复一年，人异境迁，只有两棵古榕树历尽沧桑依旧苍劲巍立。设计师将两棵古榕保留，设计中将种种承载着古榕记忆的元素拼贴，创造具有一定文化表征意义的使用空间，体现符合当地文化特质的风土人情，将植被、场地、小品等多种景观元素作为文化的载体，赋予其更多的意义，使其在发挥生态功能和使用功能的同时产生更多的文化价值。

2.3 原生景观的保留与新景观的塑造

设计中对有价值的原生植物景观予以保留，强调以本土植物造景为主来塑造具有地域特色的景观空间。大量运用大秋枫、鸡蛋花、盆架子、洋紫荆等当地乡土植物，与保留的大榕树相互辉映，营造出具有地方特色的植物景观。既节省了苗木开支，又方便日后养护，以求在节约成本的前提下达到生态效益的最大化。在硬质景观材料方面更多地选用当地加工成熟、低价的石材，灵活运用于铺地、小品及景观构筑中。

2.4 景观色谱的制定

设计构思中制定了景观色谱，方便在具体设计时控制整体景观效果，同时通过色彩语言来传达该项目特有的气质。以灰色和绿色为主调，在大尺度空间选择了淡彩（大面积的绿色植物和灰色道路铺地）来概括整体，小尺度空间则运用重彩(部分开花及色叶植物、红褐色的单体建筑、铜制雕塑、黄色平台铺地)来突出本体。厚重中带着鲜活，朴素自然，力求建筑与景观空间更好地融合，为居住区创造出宁静致远的意境空间。

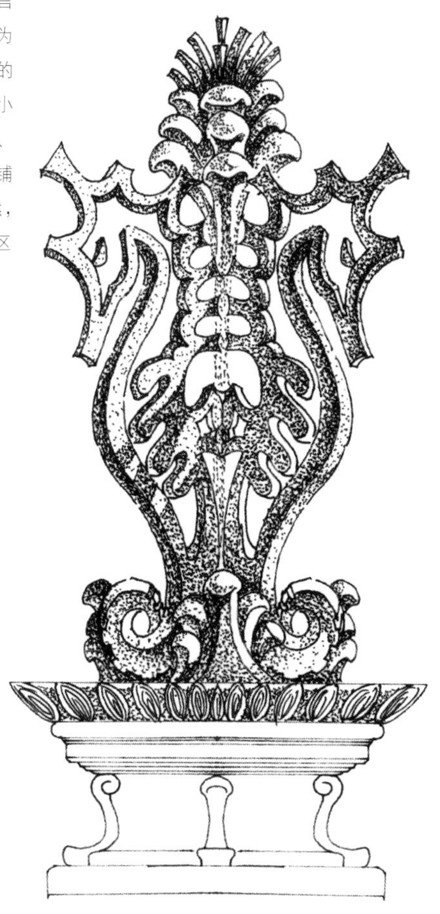

1 | 2

1 large seahorse water spray sculpture in the lake
2 manuscript of large seahorse water spray sculpture

1 湖中大型海马喷水雕塑实景
2 大型海马喷水雕塑设计手稿

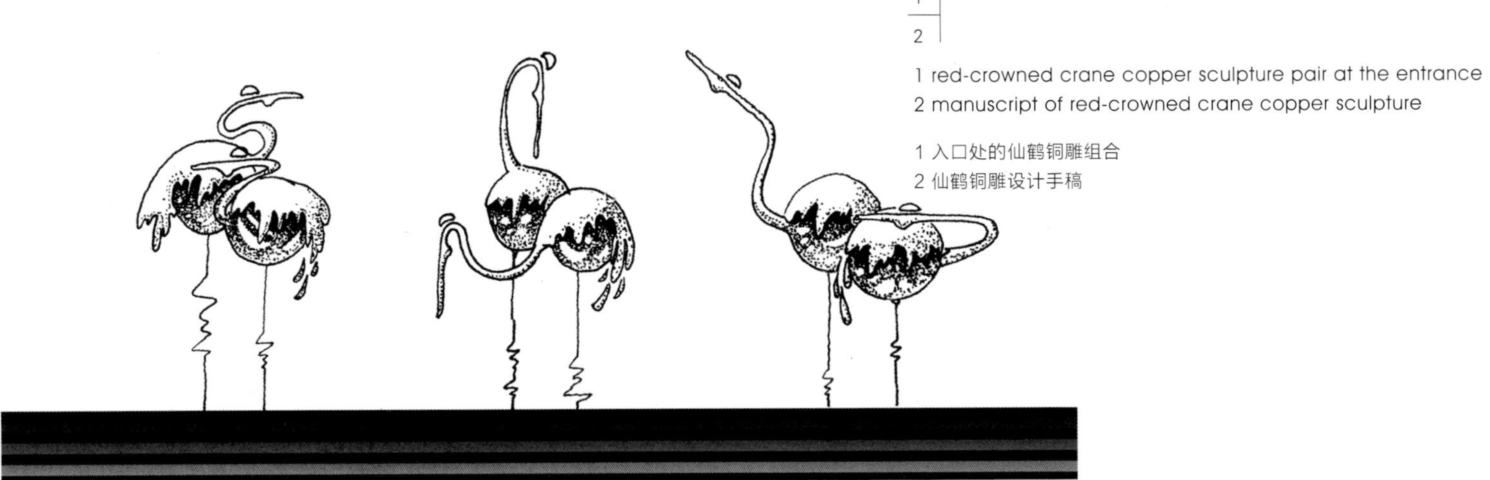

1 red-crowned crane copper sculpture pair at the entrance
2 manuscript of red-crowned crane copper sculpture

1 入口处的仙鹤铜雕组合
2 仙鹤铜雕设计手稿

2.5 景观轴线的确定

2.5.1 东西景观轴线

东西向的景观轴线为一条水轴线。超大型的人工湖成为整个居住区的焦点，湖面气势宏大，最大跨度70m，平均跨度35m，绵延长度达600m，湖面总面积约18 000m²。人工湖建造在地下车库顶板上，湖岸流畅简洁，勾勒出湖面自然流畅的曲线，水体有收有放、有源有尾。湖体与周边的植物及绿地共同形成了清晰自然的景观脉络。水边设置了临水广场、亲水平台、木栈道、溪涧、喷泉、涌泉等观水、听水、戏水的场所，营造多样化的亲水空间，给予人们心灵放松的空间。

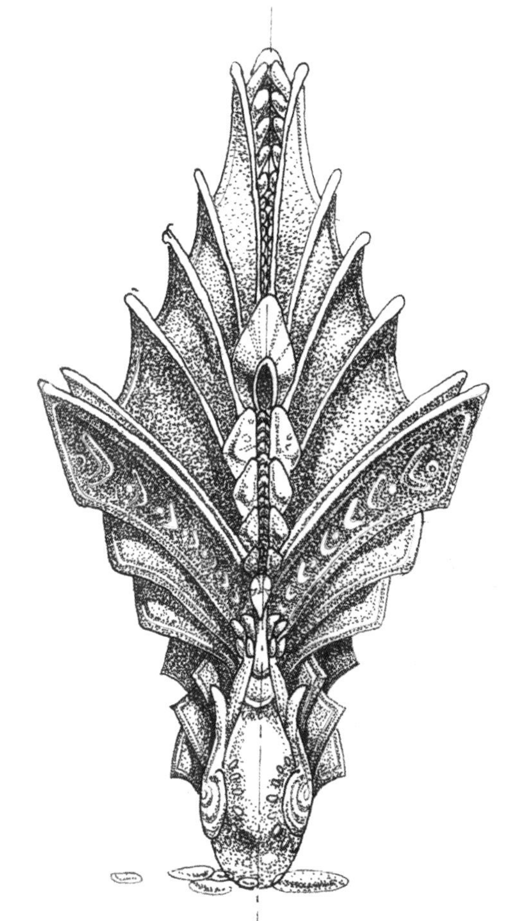

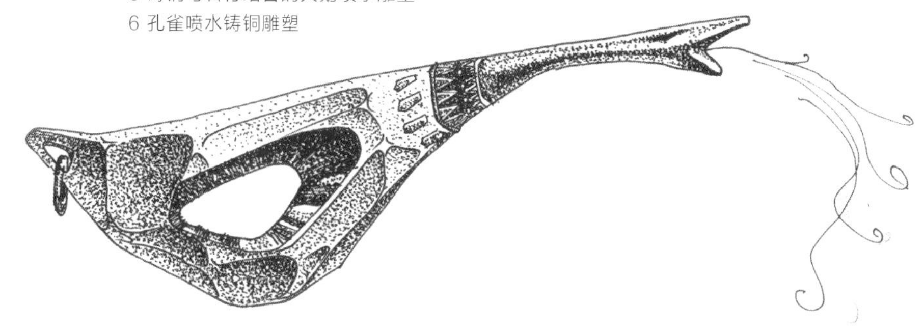

3	4
5	6

3-4 hand drawing of the sculptures
5 cast copper & stone swan water spray sculpture
6 peacock water spray cast copper sculpture

3-4 雕塑设计手稿
5 铸铜与石材结合的天鹅喷水雕塑
6 孔雀喷水铸铜雕塑

2.5.2 南北景观轴线

南北向景观轴线是一条自然艺术轴线，从南面主入口山水广场开始，跨越凌驾湖上的景观桥，与北面山体公园相接，主入口山水广场结合地下车库形成的高差将入口打造成山径、湖景、高尔夫景递进社区、渐入胜景的景观连接。宏伟的瀑布川流于层次交叠的天然山石间，数株百年古龄极品罗汉松、皇室御用黑松如游龙护珠，使入口景观既有东方园林的俊逸秀雅，又不失尊贵大气。同时，充分运用了地块的现状、地貌和植被条件，建造北后方的山体公园，以绝少的人工痕迹为居住者创造了一处具有山林野趣的休闲空间。

2.6 植物配置特点

植物配置借用了茶道“和敬清寂”的精神内涵，注重提升楼盘精神质素。大片与水相接的疏林草坪为人们提供了亲水的交流空间，大乔木以散布的种植方式将完整的草坪处理成了具有领域性、多样的小空间。在局部需要密植的地方，采用乔木、灌木与藤蔓植物结合，常绿植物和落叶植物，速生植物和慢生植物相结合的配置手法，以形成空间层次丰富、高低大小错落的自然植物群落。湖岸线以垂柳为基调树，水生植物过渡接壤，形成以花叶芦竹、风车草等植物为主的湿地景观区，恣意随性，自然而野趣。

2.7 细部景观亮点

社区围墙：800m 长的嵌铜浮雕围墙，高 4.5m，精致手绘艺术，已申请国家专利。

揽月塔：塔高 10 余米，位于园中西南，设计风格融合现代简约和古朴厚实，仰观天机瞬变，俯察万象生变，乃风水之蕴承。

艺术铜塑：应不同景观空间设计不同风格的雕塑，仙鹤雕塑为山水入口平添灵气，动物雕塑为艺术园林带来野趣，形态各异的喷水雕塑、音乐人像雕塑亦为景观营造了深远的意境。

	2-4
1	5

1 stone & cast copper squirrel sculptures in the wood
2-4 manuscript of sculptures
5 exquisite landscape wall with hand-drawing relief

1 林中石材与铸铜结合的松鼠雕塑
2-4 雕塑设计手稿
5 精致的手绘艺术嵌铜浮雕围墙

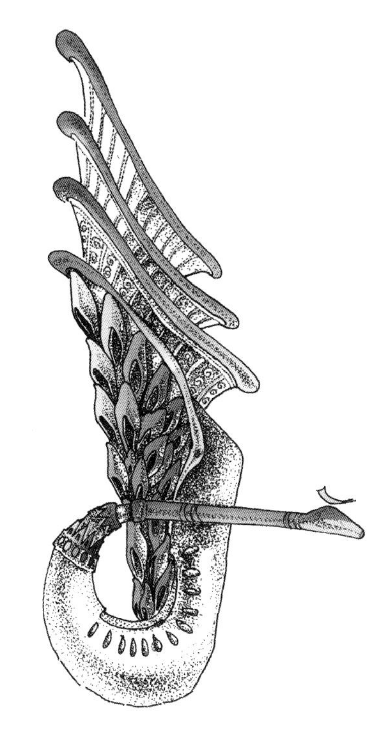

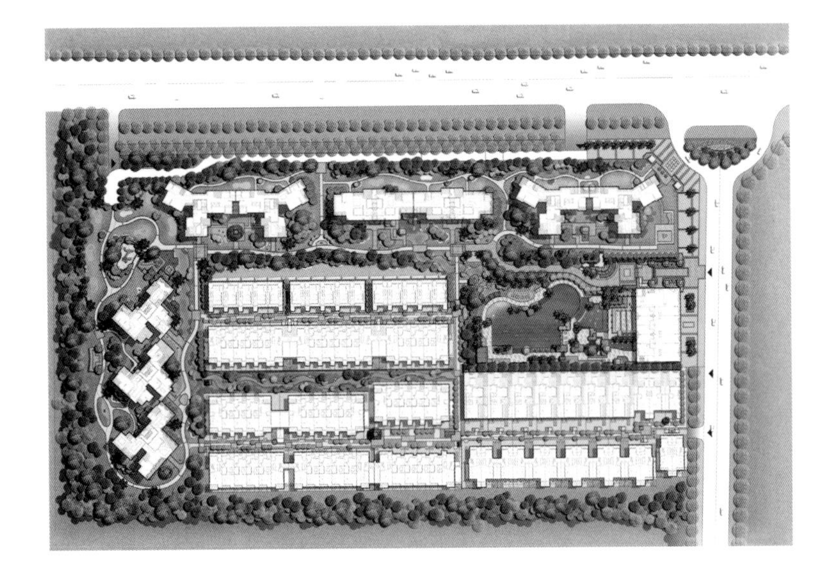

Four Seasons Garden, Dongguan
四季豪园，东莞
——领略四季风景，感悟豪情生活

Developer: Dongguan Zhongbao Four Seasons Group
Project Type: Large-scale Integrated Residential Area
Project Area: About 75,000 m^2
Design Content: Landscape Conceptual Design & Production Design
Design Period: 2008

委托单位：东莞中宝四季集团
项目类型：大型综合居住区
项目面积：约 75 000 m^2
设计内容：景观方案及施工图设计
设计时间：2008 年

The project has won the second prize of 2015 Guangdong Excellent Engineering Survey and Design; the third prize of 2015 the 5th Huacai Award (landscape design category); the first prize of 2014 Guangzhou Excellent Engineering Survey and Design (landscape design category).

该项目荣获 2015 年度广东省优秀工程勘察设计二等奖，2015 年第五届“华彩奖”园林和景观设计类铜奖，2014 年广州市优秀工程勘察设计评选园林组一等奖。

The project is located by the Fourth Ring Road in Chang'an Town, Dongguan City, Guangdong, where the traffic is convenient and natural environment is superior, overlooking Lotus Mountain on the northeast. Design philosophy: nourish heart, nourish temperament, respect native landforms, create 1,400 m^2 water fall double pools, use large area of waterscape and arrange sculptures, furniture, cascade and streams rationally. In a word, designers respect the original terrain, preserve the natural landscape and make a perfect combination of building and landscape. The semi-enclosed layout allows each household in each building to view the garden inside and mountain & water outside. The private club and the four aerial clubs, as one of the key points of the project, redefine the new standard of the good life. By concentrating the unique charm of Southeast Asian garden, it is really a high-end resort style garden community.

项目位于广东东莞长安镇四环路，交通便利，可远眺东北莲花山脉，自然环境优越。设计理念：养心、养性、养情，尊重原生地貌，1 400 m^2 跌水双泳池，大面积的水系景观运用，雕塑、小品、叠水、溪涧等合理布局，设计师尊重原有地形，保留自然风貌，让建筑与景观紧密结合，和谐共生。项目以半围合式规划，让每栋建筑户户有景，内赏园景，外观山湖，完美视野一览无遗。作为本次设计重点之一的私家会所及四大架空层泛会所，重新定义了优越生活的新标准。整个项目浓缩了新东南亚风情园林的独特魅力，堪称高品质度假风情园林大宅。

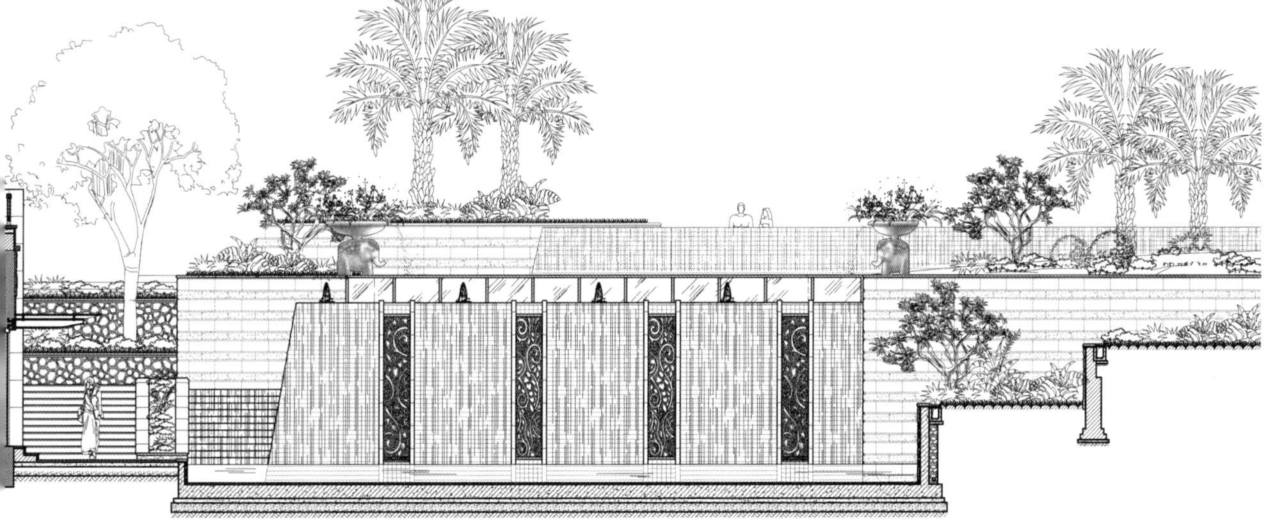

1 characteristic waterscape wall
2 elevation of waterscape wall

1 特色水景墙
2 水景墙立面图

1 | 2
  | 3

1 central garden pool area
2 elevation of Southeast Asian style pergola
3 sunken waterfront space

1 中心园林泳池区
2 东南亚风情花架立面图
3 下沉亲水空间

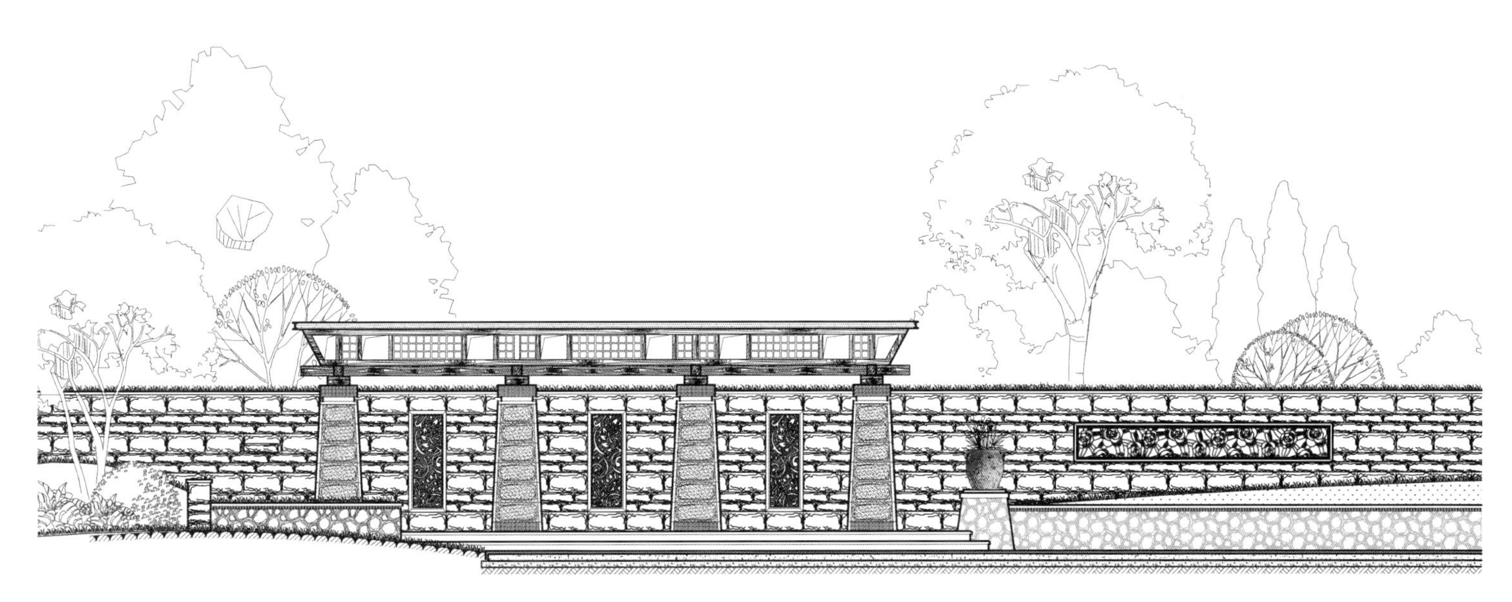

1	2	4	
3		5	6

1 gentle and quiet pool space surrounded by green lvegetation
2 blue pool and warm-colored wooden platform
3 exquisite details in a corner of the pool
4 different visual effects created by different materials, colors and volumes
5 outlet manuscripts
6 outdoor shower area

1 四周的绿色屏障围合出柔和静谧的泳池空间
2 碧蓝的泳池、暖色的木平台
3 泳池一隅精致细节
4 不同材质、色彩及体块展现出不同层次的视觉效果
5 出水口手稿
6 室外淋浴区

Four Seasons Garden, Dongguan

1 southeast Asian style viewing pavilion
2 waterfall cascade
3 diversified garden path landscape
4 pergola by the swimming pool
5 side elevation of viewing pavilion
6 simple visual effect and natural space created by the pergola

1 东南亚风格观景亭
2 瀑布跌水
3 丰富多样的园路景观
4 泳池边的休闲廊架
5 观景亭侧立面图
6 廊架营造出质朴的视觉效果和自然空间

1	3	4
2	5	

1 entrance of villa area
2 hand drawn rendering of the courtyard space that leads to villa
3 bird's eye view of entrance hall
4 landscape between villas
5 courtyard space leads to villa

1 别墅区入口
2 别墅入户的庭院空间手绘效果图
3 入户门廊鸟瞰
4 别墅宅间景观
5 别墅入户庭院空间

Everbright Beautiful Country, Dongguan
光大 · 锦绣山河，东莞
——松山湖畔，锦绣山河，独享一线湖景

Developer: Guangdong Everbright Enterprise Group
Project Type: Large-scale Integrated Residential Area
Project Area: About 3,670,000 m^2
Design Content: Landscape Conceptual Design & Production Design
Design Period: 2005

In September 2007, the project won the "Creative Prize for the Environment of Global Happy Community"; in November 2007, it was awarded "2007 Top 10 Villas of China"; in April 2008, it was honored the title of "2007 Guangdong Harmonious Community Construction Enterprise - the Most Influential Project"; in November 2008, it won the gold prize for the sustainable project in the competition for "International Garden Cities"; On July 23, 2010, it won the "2010 China Villa Jinding Award" in the 7th China Villa Fair.

委托单位：广东光大企业集团
项目类型：大型综合居住区
项目面积：约 3 670 000 m^2
设计内容：景观方案及施工图设计
设计时间：2005 年

该项目 2007 年 09 月获得"全球人居环境幸福社区创新奖"；2007 年 11 月被评为"2007 中国年度十大最佳别墅"；2008 年 04 月获"2007 广东地产龙虎榜和谐社区建设典范企业——最具城市影响项目"荣誉称号；2008 年 11 月摘得"国际花园城市"评选环境可持续发展项目类金奖；2010 年 7 月 23 日在"中国别墅十年大典"暨 2010 第七届中国别墅节中获"2010 中国别墅金鼎奖"。

Located in the core area of Songshan Lake New Town, the project has a pretty low plot ratio. It is surrounded by four colleges and universities as well as the "eight new attractions of Songhu", namely, Songhu Yanyu, horticultural expo center, Songhu Sea of Flowers, Camdor Harbour Golf Club, sunny beach, wetland park, central park and Zhuangyuanbi Park. These all combine together to create a unique eco environment. Within the site, Luqi Lake is connecting with the beautiful Songshan Lake to present a visual corridor as well as leisurely spaces for the citizens. The development features complete supporting facilities such as kindergarten, primary school, private club, star hotel, etc. It deserves the social recognition as a "top masterpiece".

Landscape Planning:

The site is divided into two parts: the project for the east part doesn't start yet, while the west part which is under construction includes a landscape avenue, five residential areas, five parks (hilltop park, lake park, lakeside park, dam park and dam crest park), hotel, club and golf court. It presents a series of sequential landscape spaces which keep harmony with the natural environment. Based on the idea of "creating a high-quality, eco and mature community", the designers have taken advantage of the existing conditions and avoid damage to the natural environment. Moreover, the design has followed the style of the

site plan of Songshan Lake 松山湖规划总平面

surroundings and brought new vitality to the community to highlight its elegance.

Key Landscapes in the West:

West Entrance: the development starts with majestic water features at the west entrance. On the left, rocks and cascades are metaphors of hills and rivers, which implies the magnificence of the "Beautiful Country"; on the right, water feature is designed with rough stone facing, looking elegant and luxurious.

Landscape Avenue: a broad landscape avenue runs through the west area from south to north, connecting different building groups together. The designers set a platform to indicate the entry; other landscapes are designed according to the terrain. All these efforts are made to create a natural, ecological and comfortable landscape avenue.

Lake Park: the design has retained the existing shoreline and upgraded the landscape environment. Garden paths are built according to the terrain and keep contact with the lake to provide unique strolling experience.

Lakeside Park: The design shows great respect to the original terrain and upgrades vegetation to diversify the lakeside landscape spaces.

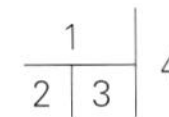

1 4,000 m² infinite swimming pool
2 enjoying unparallel waterscape
3 natural and elegant curved line around the swimming pool
4 entrance lobby of the club

1 4 000 m² 无边泳池
2 坐享无敌深邃水景
3 泳池边有着自然简约的曲线
4 泛会所主入口大堂

光大·锦绣山河位于东莞松山湖新城核心地段，拥有超低容积率。锦绣山河周边有大学院校4所，并有号称松湖新八景的松湖烟雨、园艺博览中心、松湖花海、金多港高尔夫、阳光沙滩、湿地公园、中心公园、状元笔公园，形成了独有的自然生态景观氛围。项目内的鹭栖湖和风光旖旎的松山湖相连接，既能形成景观视线通廊，又为周边地区居民提供了良好的休闲空间。园区配套设施完善，除了小学和幼儿园，也有私人会所和星级酒店等休闲娱乐配套。无论从规模、配套、交通还是园林景观，该项目都堪称珠江东岸第一盘，是当之无愧的塔尖之作。

景观构架

整个园区分为东、西两大片区。东区地块尚未启动，目前在建的西区内规划有景观大道、5个居住区、5大公园（山顶公园、湖边公园、湖滨公园、大坝公园、坝顶公园）、酒店会所及高尔夫练习场。社区规划呈现出疏密有致的景观空间，各区有机排布，空间收放自如，和谐有序，浑然天成。整体景观设计以“创造高品质的自然生态成熟社区”为理念，以自然的视角，因地造景，避免人工匠气，在延续松山湖板块的整体城市风格基础上，赋予其新的生命活力。

西区重点景观

西入口：以磅礴大气的水景为开篇画卷，左侧景观以大石喻山、以跌水喻河，象征“锦绣山河”之壮美，右侧水景则以粗犷石材饰面渲染出一份大气与华贵。

景观大道：宽阔的自然景观大道贯穿西区全园南北，将区内各组团连接起来。设计师在入口处以节点平台作为提示作用，在其他区域则以地形绿化为主要景观元素，着力打造生态自然、舒适高档的小区景观大道。

湖边公园：设计中保留了原有的湖岸线，在最大化利用原有资源的基础上进行局部优化改造。园路与地形相结合，并与水面维持若即若离的空间关系，增加了人在其中游走的趣味性。

湖滨公园：景观设计充分尊重原地形，主要在植物配置上进行优化改造，使湖岸绿化层次更加丰富，种类更具多样性。

1	4
2 | 3	5

1 landscape at the main entrance of Plot 4
2 garden gate at the entrance
3 open regular lawn
4 interior of the main entrance lobby
5 elevation of the main entrance lobby

1 4 号地块主入口前景
2 入口园门
3 开放式规则草坪
4 主入口大堂内庭
5 主入口大堂立面图

Everbright Beautiful Country, Dongguan

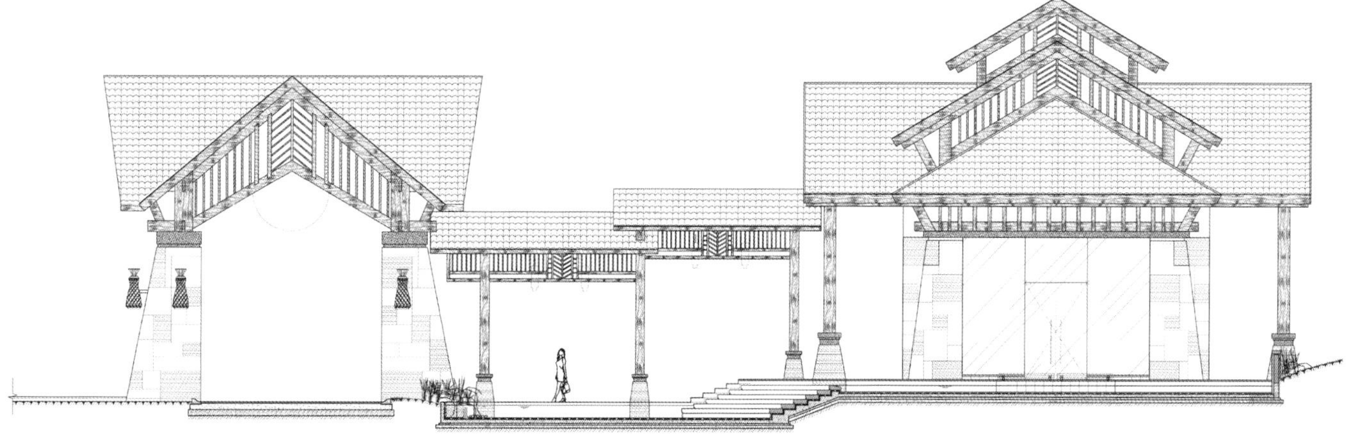

1/2 | 3

1 hand drawing of the sculptures
2 cast copper zebra sculptures on the lawn
3 cascade formed by cyan slates

1 雕塑手稿
2 草坪上的铸铜斑马雕塑
3 利用高差构成的跌水由莹青板砌成

1 | 2
  | 3

1 surprising landscapes behind the thick forest
2 elegant cascades
3 details of the lakeside lamps

1 从茂密的树林望出去，总能发现惊喜
2 错落有致的叠水空间
3 湖边灯具细节

1	3	4
2	5	6

1 elegant crane sculpture
2 bronze peacock sculpture
3 hand drawing of the crane sculpture
4 hand drawing of the peacock sculpture
5 bronze peacock sculpture
6 big bird sculpture made of bronze and stone

1 姿态婀娜的仙鹤雕塑
2 青铜孔雀雕塑
3 孔雀雕塑手稿
4 仙鹤雕塑手稿
5 青铜孔雀雕塑
6 铸铜与石材相结合的大鸟雕塑

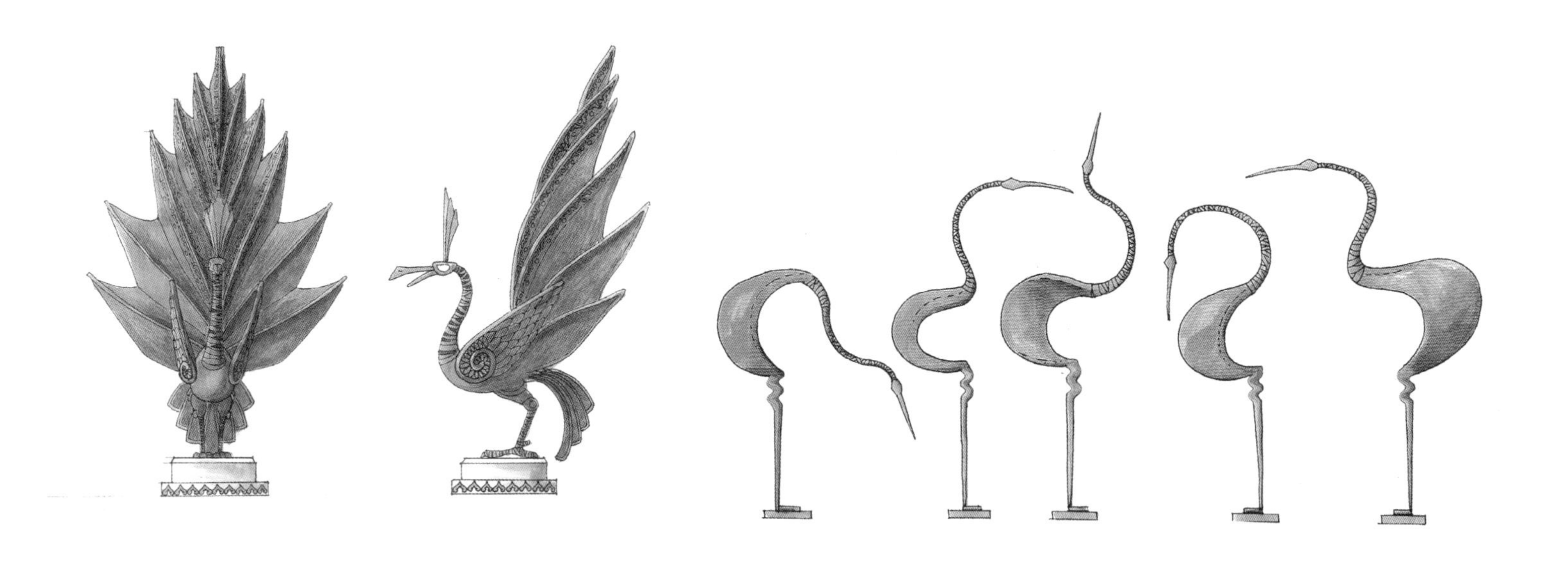

1 | 3
2 | 4

1 hand drawing of deer sculptures
2 lively deer sculptures in the forest
3 flamingos standing by the rocks and cascade
4 hand drawing

1 鹿群雕塑手稿
2 丛林中栩栩如生的鹿群
3 叠石跌水旁站立着几只体态轻盈的“火烈鸟”
4 手绘效果图

Vanke Dream Town · Ming, Guangzhou
万科城 · 明，广州
——东方院墅，庭院深深

Developer: Vanke Estate Group
Project Type: Low-density Residential Area(Villa Area)
Project Area: 28,250 m^2
Design Content: Landscape Conceptual Design & Production Design
Design Period: 2008

委托单位：万科地产集团
项目类型：低密度居住区（别墅区）
项目面积：28 250 m^2
设计内容：景观方案及施工图设计
设计时间：2008 年

The project was titled "Gold Villa" in 2009.
该项目获 2009 年度金牌别墅称号。

1.Surrounding Environment

Vanke City is located in the eastern part of Guangzhou, at the intersection of Kaichuang North Road and Guangshan Road, and borders Science city on the north. Since the land is originally the location of overseas Chinese farm, a variety of rare species are retained, especially the 70+ hundred years old lychee tress. Leaning against the mountain, it boasts beautiful environment, broad vision and exclusive leisure and elegance.

The third phase are modern Chinese courtyard villas located in the southeast of Vanke City, adjacent to two main roads: Guangshan Road and Kaichuang Road. With the high rises as its screen, the villa area enjoys a great residential pattern — "downtown outside, secluded residence inside".

2.Landscape Culture & Design Philosophy

Through analysis and consideration, designers come up with a new idea to create an ideal residential landscape: it is necessary to have the aesthetic taste of modern art and to fully excavate local culture, via suggestive practices to raise cultural awareness of traditional neighborhood, embodies the perfect combination of art and function, as well as the perfect combination of traditional living pattern and modern landscape aesthetics. Therefore, the project focuses on the Guangzhou Lingnan culture, use landscape wall and height different to create the sight of winding path leading to a secluded spot in traditional garden, highlight the affectionate Fengshui design and forming a landscape pattern of "rich family in spacious courtyard".

3.Theme of Space

"Vanke Dream Town·Ming, Guangzhou" is divided into different theme spaces which are characterized differently and get on well with each other as an organic whole.

Greenway space: the greenway planted with big-crown trees forms a distinctive entrance area. Dominated by soft plants, the pleasant and private green space becomes the green background for the entrance area.

Garden space: the garden space combines with greenway to form the "sitting room" of the community. Terraced structures and landscape walls create varied alleys outside the buildings and in the central courtyard a sunken space is

1-4 Landscape between villas
1–4 别墅宅间景观

formed naturally. In addition, the pool area is the largest landscape node which is uplifted in accordance with the terrain, creating a dynamic flowing cascading water-wall.

Alley space: enclosure landscape wall and plants are important in this area. Alley is not just functional passageway, but a link to connect node spaces in different scales for different uses, thus to form rhythmic landscape.

High-rise space: high-rise area emphasizes transparency and continuation of sight, and it is important to link the ground empty space with outdoor space. Height difference is used to form a series of landscape waterways, forming another "water alley" layout with garden path. Meanwhile, well-arranged plants are integrated into interior space to create continuous landscape. Corridors and scenic spots are set in the empty space to welcome "cross ventilation", what a pleasant transitional landscape space.

4.Way Back Home

Pedestrian flow and vehicle flow are separated, which is still rare in the low-density townhouses, to provide residents a living experience belonging to traditional luxury courtyard. In addition, the separation maintains the landscape integrity, reduces the pollution caused by automobile exhaust and guarantees the safety of the young and the old.

5.Plant Design

In order to outline a clear and relaxed oriental prospect, designers propose to act in accordance with architectural culture. Plenty of black bamboos are planted by the winding secluded lanes and high walls, which bring out the bright and clean dust-free buildings and landscapes and highlight the elegance and tranquility of Chinese garden. In addition, greenery is dotted with frangipani, bougainvillea, Gomphrena globosa, Yucca aloifolia, swartzia and so on, presenting a wonderful space that is rich in color and fragrance.

1 | 2
  | 3-6

1 small garden at the entrance of the villa
2 elevation of the enclosing wall
3 entrance of the villa
4 backyard
5 inner garden
6 enclosing wall

1 别墅入口前庭院
2 别墅院墙立面图
3 别墅大门入口
4 后院
5 内院
6 围墙

一、周边环境

万科城位于广州东部新兴的豪宅板块，科学城北部，开创大道北与广汕路交界处。地块原为华侨农场所在地，故项目现场保留多种珍稀树种，以70多棵逾百年历史的荔枝树尤为珍贵；项目依山而建，环境优美，视野开阔，独享清闲与雅致。

三期组团“万科城·明”现代中式院墅，地处万科城东南方，坐拥整个项目最佳地理位置，本案邻近两大主干道：广汕公路和开创大道，借高层区为屏障，使别墅区形成“出则闹市，入则幽居”的绝佳居住格局。

二、景观文化和设计理念

通过对“万科城·明”的分析与思考，我们对营造理想居住景观有了新设想：既要具备现代艺术美学品位，又要充分挖掘本地乡土文化气息，通过形式暗示的手法，唤起人们对传统街区小巷的文化记忆，体现艺术与功能的完美结合，同时也是传统居住行为模式和现代园林美学法则的完美结合。因此，设计将项目聚焦在广州地区特有的岭南文化，运用景墙与场地高差变化形成传统园林中的曲径通幽的意境，突出曲折有情的风水设计，形成“宅院深深，大户人家”的景观格局。

三、空间主题

“万科城·明”划分为若干不同的主题空间，主题空间各具特色，同时又相互融和，成为一个有机的整体。

林荫道空间：入口林荫景观道形成风格鲜明的小区入口，以展示性的景观为主导。林荫景观道，以植物软景为主景，营造舒适宜人、密闭性较强的绿荫空间，同时成为入口区的绿化背景。

花园空间：与林荫景观道结合形成“先抑后扬”的景观格局，兼具小区客厅的功能。通过阶梯构筑物和景墙在建筑外围形成虚实变化的小巷，中央庭院形成一个凹陷的围合性空间，如浑然

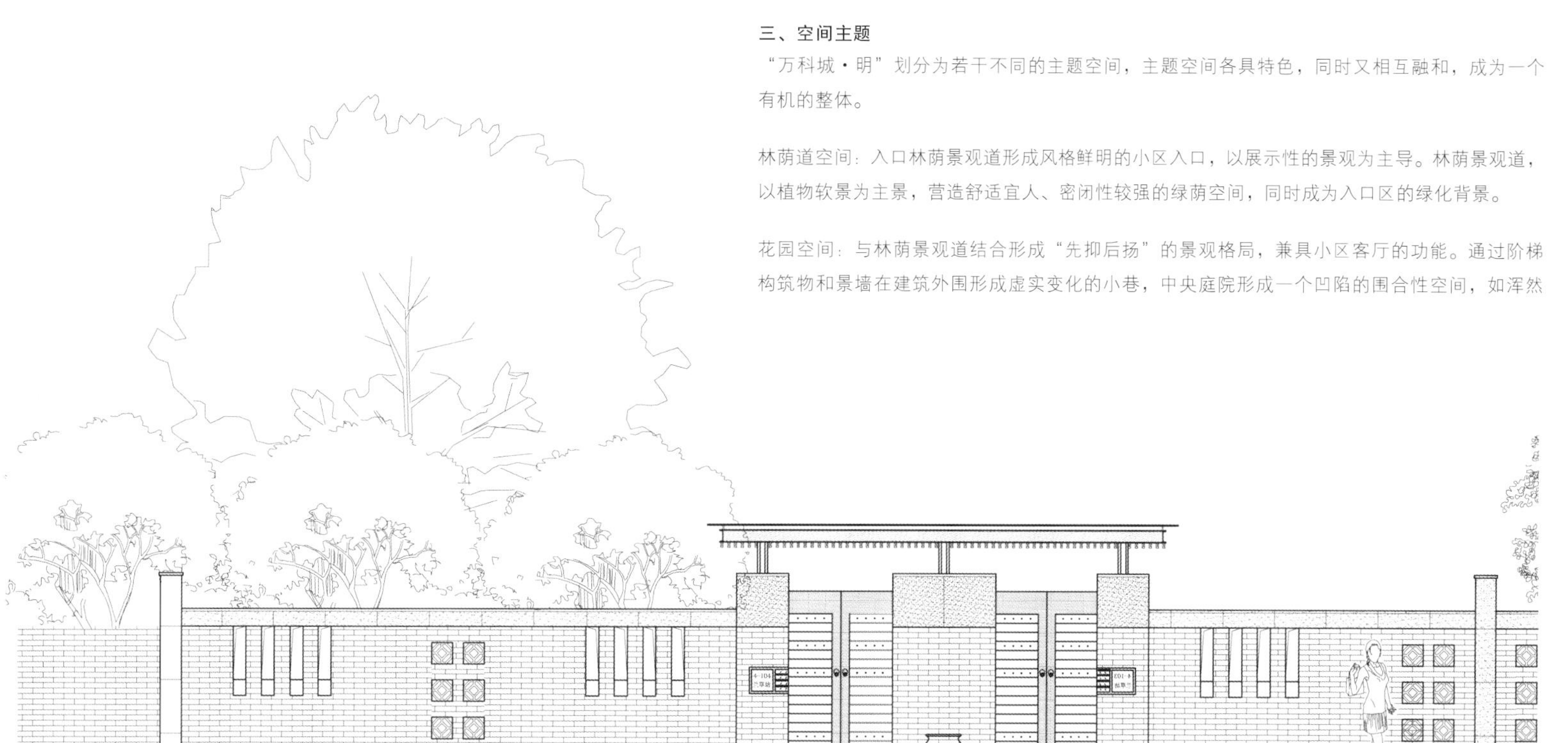

1 harmony between architecture and landscape
2 bougainvillea and plumeria in the bamboo forest
3 elevation of the landscape wall at the entrance of the lane
4 colorful vegetation at the entrance of villa area

1 建筑与景观和谐相融
2 翠竹中点缀着勒杜鹃和鸡蛋花
3 胡同入口景墙立面图
4 别墅区入口丰富的植物配置

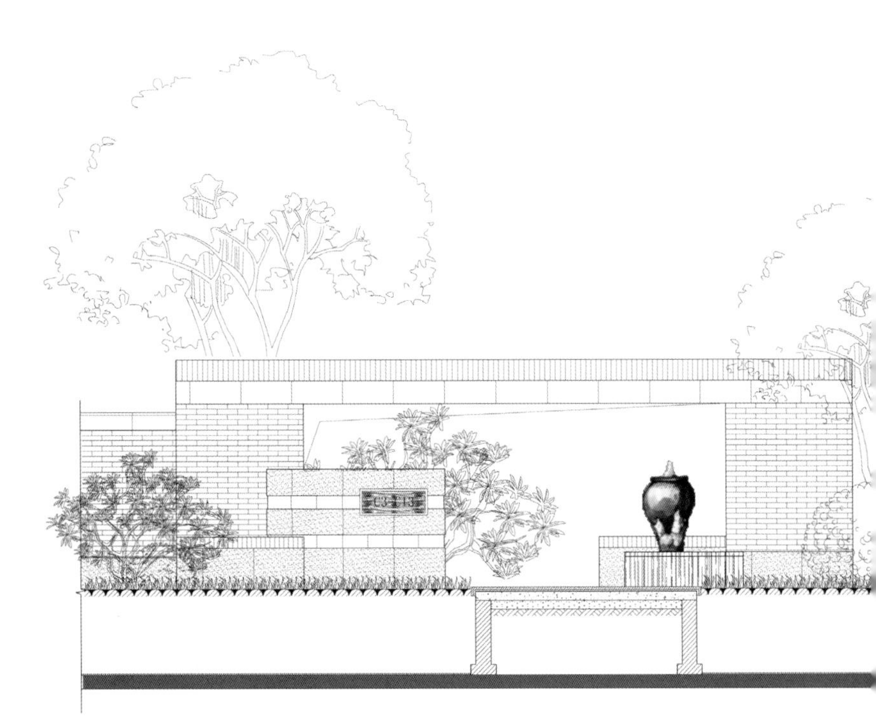

天成。另外，泳池区是最大的景观节点，顺应地形作抬升处理，形成叠级水墙“动”景的效果。

胡同空间：围合的立面景墙以及植物的运用是此区的重点。胡同、小巷既是功能性的通道空间，也会串连一些尺度各异的不同性质的节点空间，形成变化富有节奏韵律的景观效果。

高层区空间：高层区强调视线的通透、延续，架空层与室外空间的衔接是景观的重点。运用高差形成系列景观水道，与景观园路相结合形成另一种“水巷”的布局，结合植物的围合与间隔，与室内空间形成“形断意连”的景观效果。架空层利用柱、墙所分隔的空间，设置廊道与观景点，组织“穿堂风”形成舒适宜人的室内过渡性景观空间。

四、归家线路

结合车行与步行路线组织，采用了人车分流的设计，这在低密度联排住宅中尚不多见。通过人车分流，组团可顺利营造出巷陌交错、高墙大院的居住感受。同时，人车分流维持了园林的完整性，减少了汽车尾气带来的环境污染，也保障了儿童和老人的居住安全。

五、植物配置

为了勾勒出万科城·明淡定从容的东方意境，我们从大景观的角度出发，提出与建筑和文化相结合的植物配置方案。在曲曲折折的幽巷高墙旁，遍种紫竹，衬托出建筑和景观的明净无尘，凸显出中式园林清雅、恬静的基调。在成片的绿色中，还点缀着鸡蛋花、勒杜鹃、千日红等开花植物和七彩马尾铁、红铁等色叶植物，花香四溢中又增添了色彩的丰富变化。

1	3	4	5
2	6	7	

1 elevation of swimming pool area
2 landscape at the entrance of the villa area
3-5 form cascade landscape through the height difference between the swimming pool and the ground
6 swimming pool lounge chair leisure space
7 children's swimming pool landscape

1 泳池区立面图
2 别墅入口景观
3-5 利用泳池与路面的高差形成跌水景观
6 泳池躺椅休闲空间
7 儿童泳池景观

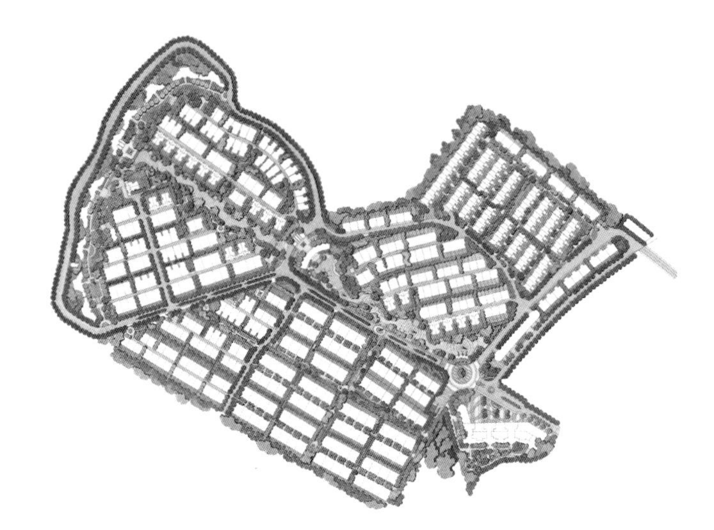

1
2 | 3

1 fountains in the lake
2 partly hidden and partly visible villas
3 cascades at the entrance

1 自然式湖面喷泉
2 别墅若隐若现
3 入口特色跌水

Agile Flowing Garden, Guangzhou
雅居乐·小院流溪，广州
——倚山、近水，诗意栖居

Developer: Agile Property Holding Limited
Project Type: Large-Scale Integrated Residential Area
Project Area: 442,866 m²
Design Content: Landscape Conceptual Design & Production Design
Design Period: 2010

委托单位：雅居乐地产控股有限公司
项目类型：大型综合居住区
项目面积：442 866 m²
设计内容：景观方案及施工图设计
设计时间：2010 年

Located next to Liuxi River in the west and leaning against hills in the east, the project enjoys advantaged natural resources and complete supporting facilities, i.e. schools, hospitals and supermarkets etc. Following the ideas of creating "a waterfront eco community for health and vacation", the designers has introduced the water of Liuxi River to build a waterscape and a viewing platform in the center. At the same time, static waterscape combines with dynamic water features to present colorful landscape effect. Modern art sculptures and landscape architectures are set to highlight the modern and poetic atmosphere of the community.

该项目西临流溪河，东倚山地丘陵，自然环境得天独厚且周边已有成熟社区，并有学校、医院、超市等各种资源。遵循“养生度假的水岸生态社区”的定位，在总体设计中，设计师充分利用了近流溪河水的优越性，引水打造区内的中心休闲水带，并设置观景平台，将水的观赏性和实用性得到了最大限度的发挥。同时，在中心休闲水带的营造中，平静典雅的静态水结合充满朝气的动态水，形式多样，为社区增添了多变的景致。通过在最佳观景点点缀具有现代感的艺术雕塑、小品等构筑物，突出社区现代、诗意的栖居气质。

小院流溪
GARDEN

沐溪

	1	4
2	3	

1 landscape architectures beside the lake
2 cascade and fountain at the entrance
3 garden path leading to home
4 pergola

1 湖边情景小品组合
2 入口涌泉跌水
3 归家园路
4 休闲廊架

1 | 2 | 3

1 big swimming pool of the club
2 water bowl
3 water features and lamps

1 会所大泳池
2 特色景观水钵
3 水景灯具

Financial Street Rongsuihuafu , Guangzhou
金融街 · 融穗华府， 广州
——由繁入简，有舍有得

Developer: Financial Street Holdings
Project Type: Integrated Residential Area
Project Area: 25,000 m^2
Design Content: Landscape Conceptual Design & Production Design
Design Period: 2015

委托单位：金融街控股
项目类型：综合居住区
项目面积：25 000 m^2
设计内容：景观方案及施工图设计
设计时间：2015 年

Located in the north of Guanggang New Town, Liwan District of Guangzhou, the development borders Guanggang Central Park on the south, enjoying complete facilities and convenient traffic.

The overall layout makes reference to the natural contours of Baiyun Mountain and Pearl River. Grassy slopes and cascades imitate the entrance of mountain, and the garden path covered by fallen leaves leads people to the open relics plaza. Natural space design presents an atmosphere of fairy land.

Since the site was for industrial use, the design has fully respect the history and retained some industrial relics to create funny landscapes together with green plants. In this way, it has built a modern cultural residential environment with historical memories.

金融街 · 融穗华府位于广州市荔湾区广钢新城北部，南接广钢中央公园，尽享核心配套，周边交通十分便利。

项目借用白云山与珠江的自然轮廓，抽象概括为场地布局线条。景观伊始是草坡结合特色跌水，营造云山蔼蔼、流水潺潺的入山意象，经过落英缤纷的林荫通道巧妙抑景、酝酿情绪。视线在林荫道后的遗址广场空间得以放开，充分提升空间感受的品质感，收放自如间颇有桃源氛围。

场地原址拥有悠久的历史工业记忆，设计中充分尊重场地精神。保留工业遗址，配合精致绿化，突出场地文化符号的同时使之更有园林趣味。营造出一处既饱含刚强的历史余韵、又充满柔美的文脉气息的现代人文生活场所。

1 multi-level landscape at the entrance
2 water pool and landscape wall

1 入口造景层次丰富
2 水池景墙别具创意

1	2	5
3	4	6

1-2 flowers soften the hard rocks
3 violin-shaped flower boxes
4 double-level phytocoenosium along the road
5 early morning light wakes all nature
6 financial street in the night

1-2 鲜花柔化了片石的冷硬
3 提琴形花箱
4 路旁的复层植物群落
5 晨光里，生命勃发
6 夜色中的金融街

1 | 3
2 | 4

1 bird's-eye view of the children's playground
2 night view of the children's playground
3-4 happy children in the playground

1 儿童活动区鸟瞰
2 儿童活动区夜景
3-4 乐园里的欢乐背影

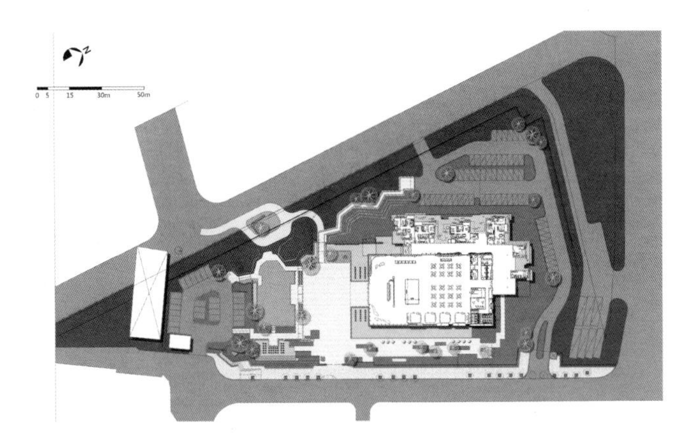

Yuexiu Starry Haizhu Bay, Guangzhou
越秀 · 星汇海珠湾展示区，广州
——云山珠水，纸承匠心

Developer: Yuexiu Group
Project Type: Integrated Residential Area
Project Area: 17,585 m^2
Design Content: Landscape Conceptual Design & Production Design
Design Period: 2015

委托单位：越秀集团
项目类型：综合居住区
项目面积：17 585 m^2
设计内容：景观方案及施工图设计
设计时间：2015 年

Located in the west of Haizhu District, Yuexiu Starry Haizhu Bay sits to the opposite of GISE New City, occupying a total area of 2,450,000 m^2. As the key project in Guangzhou's strategy of "suppressing the second industry and developing the third industry", it was renovated from an abandoned papermaking factory, to be a new residential community which will accommodate 120,000 people. In addition to the waterscapes and the green lands, the exhibition area will be designed with complete traffic network, more public facilities and historical industrial scenes to create a beautiful and comfortable living environment.

Taking the surrounding environment into consideration, it preserves the big trees and uses broken lines and straight lines to create simple and elegant spaces. Yellow-grey stone façade keeps in harmony with the buildings, and the flower beds and landscape walls made of weather-proof steel plates will remind people of the industrial history of the site.

项目位于海珠区西部，面积约 2.45 平方公里，与广钢新城隔江相望。它是广州市实施"退二进三"策略的重点项目，是一个废弃造纸旧厂房改造项目，将成为一个容纳 12 万人居住的新社区。该地块将结合水景绿地，完善交通路网，增加公建配套，保留工业历史场景，打造为水秀花香的宜居城区。

简洁大气的空间设计，充分考虑场地周边的环境，对原有大树予以保留利用，平面构图上运用现代折线加直线元素，立面材质上除了整体跟建筑呼应的黄灰调石材外，特别采用了耐候钢板这一富有工业味道的材料，用在造型花基和标识景墙上，回应场地原有历史记忆。

1 black-white-grey linear pavement highlights the entrance area
2 reflection pool and fountains at the entrance of the sales center
3 night view of the entrance of the sales center

1 黑、灰、白三色线性铺装强调入口的引导性
2 镜面水涌泉为售楼部入口增添氛围
3 售楼部入口夜景

1 | 2
3

1 flower beds made of stones are in contrast to those made of weather-proof steel
2 artificial grass rolls used in landscape walls and ground pavements
3 weather-proof steel flower beds are carved with Chinese characters to remind people of the former papermaking factory

1 石材花基与耐候钢花基既在形式上统一又在材质上对比
2 仿真草卷在景墙立面和地面铺装的运用
3 耐候钢花基上镂空刻着代表纸厂记忆的文字

Yuexiu Starry Haizhu Bay, Guangzhou

4 sales center under the setting sun
5 unified broken lines create multi-level spaces

4 夕阳下的售楼部
5 统一的折线元素塑造出层级丰富的空间

1	2	5
3	4	

1 entrance of the backyard enclosed by landscape walls
2 night view of the reflection pool
3 details of the landscape wall
4 sculptures on the lawn
5 the reflection pool dialogues with the buildings

1 景墙的围合形成后场空间的入口
2 售楼部镜面水夜景
3 景墙细节
4 草坪上的情景雕塑
5 镜面水与建筑相衬相融

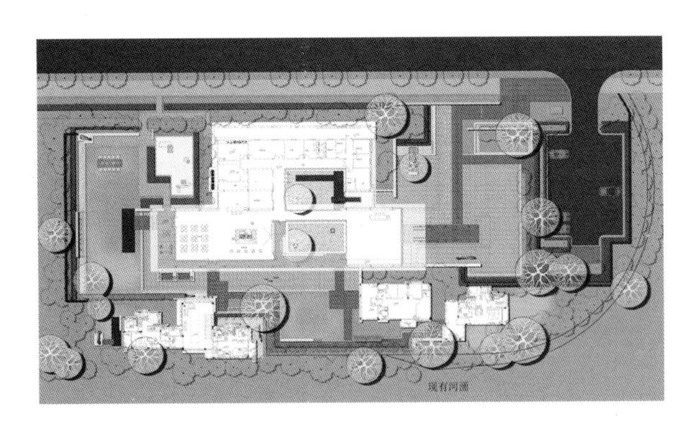

Poly i Cube, Guangzhou
保利 i 立方，广州
——沉思中心，素说禅意

Developer: Poly Real Estate Group
Project Type: Integrated Residential Area
Project Area: 69,946 m^2
Design Content: Landscape Conceptual Design & Production Design
Design Period: 2016

委托单位：保利地产集团
项目类型：综合居住区
项目面积：69 946 m^2
设计内容：景观方案及施工图设计
设计时间：2016 年

Zengcheng Poly i Cube is located on the south of Yongning Avenue, the arterial road that connects east Guangzhou with Huizhou, Dongguan and Shenzhen. The Meditation Center, as the exhibition area of Poly i Cube, is designed in modern zen style with hills, stones and water features. Inspired by Buddhism and tea culture, the landscape spaces are arranged in T shape to keep harmonious with surrounding buildings. Dominated by the magnificent mirror water, the front yard is skillfully designed to make the hills, buildings and water features dialogue with each other. Glass wall of the corridor allows people to enjoy the spacious sunshine lawn outside, and trees of different heights are employed to imitate the great view of "rising hills". In the interior meeting area, shadow of the trees floats on the sculptural landscape wall, interpreting the lifestyle of "appreciating mountains and hearing birds' song". The zen-style meditation center caters to modern people who want to escape from the mundane world. By employing simple and elegant design skills, it has created a ideal place where people can experience a natural and slow life.

增城保利 i 立方位于广州东进走廊与穗莞深大动脉的永宁大道南侧，示范区“沉思中心”以简洁精炼的山、石、水等景观元素，营造极具现代感，寄情于山水的禅意空间。提炼“禅茶一味”的精髓，将整体空间呈“品”字形布局，与建筑主体融为一体。气势磅礴的镜面水景，通过巧妙的三维一体构图，将山石与水，建筑与水，建筑与山石相互映衬，相互交融，打造极简极致前场。建筑内廊的落地玻璃将外部宽阔的阳光草坪尽收眼底，引入“层山叠嶂”的理念，通过极简的设计处理，将高低错落的树池和桩景大乔巧妙地构建侧院景观的视觉焦点。室内洽谈区的对景景墙，树影似轻纱浮动，掩映于雕塑景墙之上，书写幽寂“观山听鸟鸣”之禅意后场。沉思中心的“禅”迎合人们对远离浮华的期许，整体设计以极简的设计手法提炼山水而造势，追求城市慢生活的返璞归真，意犹未尽。

1 simple and elegant welcoming space
2 welcoming water features
3 quiet and clean garden full of flowers

1 简约而意境深远的迎宾空间
2 迎宾水景
3 静谧无尘，华满芳庭

1	2	
3	4	5

1 interesting cube
2 spacious backyard
3 drinking tea and chatting with friends
4 zen-style stone decoration
5 trees and shadows

1 有趣的立方
2 舒朗后庭
3 悦茶叙谊
4 禅意置石
5 素禅之意，枝影斑驳

6 | 7 | 8

6 artificial hills create a welcoming atmosphere
7 mirror pool and grass steps
8 experience zen style through every stone and every leave

6 层山叠峰，迎宾之境
7 镜面水景与层级草阶构筑线性美感
8 一石一叶有禅思

保利 i 立方，广州

1	2	5
3	4	6

1 boundless space with lights and shadows
2 quiet atrium under warm and soft light
3 reflections of building facades on the water
4 long corridor connects with the atrium to extend the space
5 quiet atrium space for relaxation of body and mind
6 long and quiet corridor

1 灯影相映，无界空间
2 禅静中庭灯光温馨平和
3 建筑立面线条倒映水中，韵味悠长
4 走廊中庭空间渗透，纵深之美
5 中庭静享光色，放空身心
6 悠长走廊

Poly Central Park, Heshan
保利中央公园，鹤山
——在华光掠影中，品味慢生活

Developer: Poly Real Estate Group
Project Type: Integrated Residential Area
Project Area: 159,259 m^2
Design Content: Landscape Conceptual Design & Production Design, Integration of Soft Landscape Design and Construction
Design Period: 2015

委托单位：保利地产集团
项目类型：综合居住区
项目面积：159 259 m^2
设计内容：景观方案及施工图设计，景观软装设计施工一体化
设计时间：2015 年

Located in the new center of Heshan City, on the south-north cultural axis, Poly Central Park is adjacent to Dayanshan Park and Heshan Middle School, enjoying incomparable landscape and cultural resources.

The landscape design follows the architectural style to be simple and elegant. Natural flowing lines are employed to create big bright spaces and small quiet spaces. With emphasis on culture and function, quality and visual effect, it has built an eco garden in the busy city with natural landscapes and water features. The artistic landscape architectures and features also show great care to people. Epoxy resin runway extends through the garden, thus people can enjoy beautiful landscapes while jogging. Modern design skills are used to interpret ideal Chinese lifestyle, creating an eco residential community which caters to modern living habits and features Chinese cultural atmosphere.

鹤山保利中央公园地处鹤山新城市中心，坐落于新城南北文化轴上，毗邻大雁山公园与鹤山中学，拥有绝佳的景观和文化资源。

场地建筑风格线条简约，景观营造上也聚焦体现大气时尚。运用自然流畅的大线条布局构图，巧妙营造干净疏朗的大空间与舒适静谧的小空间。通过自然山林营造与多处特色水景点睛，重点打造“引山·藏水·聚气”的城市意境山水景观。从而形成文化与功能并重、品质与视觉双赢的城市花园。园林小品、园建等各类设施在追求艺术特色的基础上充分体现人文关怀。

三级环氧健身系统跑道贯串全园，让步伐随园内山水自由起伏，身心充分感受自然的温度。

以西方成熟的现代设计语言阐述理想的中式生活理念，呈现既符合现代人生活习惯，又有中式文化韵味的人文生态社区。

1 bird's-eye view in the night
2 sales center in the dusk

1 夜景鸟瞰图
2 暮色里的售楼处

1 | 3
2 |

1 isolated tree enhances the atmosphere for meditation
2 lighting design for the cascade
3 bright lights and glittering water

1 孤植大树渲染沉思气氛
2 跌级水景灯光设计
3 灯火煌煌，水光潋滟

保利中央公园，鹤山

1	2	5
3	4	6

1 lights dialogue with fountain
2-3 simple and elegant lamps
4 lighting, beyond vanity and returning to peace
5 light, shadow and glittering water
6 logo at the entrance in the light

1 灯具与涌泉相映成趣
2-3 极简的造型灯具
4 灯光超越浮华，回归宁静的禅意
5 波光潋影，光影斑斓
6 夜色中的入口 logo 标识

| 1 | 4 |
| 2 | 3 | 5 |

1-3 delicate fabric decorations
4 logo at the entrance
5 details of the fabric decorations

1-3 精致的软装布品
4 入口 logo 标识
5 软装布品近景

保利中央公园
CENTRAL PARK

Bar
TELEPHONE
TELEPHONE
WELCOME
TO THE
LONDON
PARIS

The Orchidland, Huizhou
悠兰山，惠州
——新城市中心的顶级隐幽静谧之地

Developer: Si Fang Industrial Co., Ltd.
Project Type: Integrated Residential Area
Project Area: 94,500 m^2
Design Content: Landscape Conceptual Design & Production Design
Design Period: 2009

委托单位：四方实业有限公司
项目类型：综合居住区
景观面积：94 500 m^2
设计内容：景观方案及施工图设计
设计时间：2009 年

The project is located by the riverside ecological corridor and stands above the new city center, which blend dignified and elegant European architectural style with traditional Chinese living philosophy, creating a unique residential community in the mountain region. The building abuts against the mountain side in the sun, occupies the commanding height and overlooks the city and Dongjiang River. Within 10 minutes twin-city circle and PRD 1 hour living area, it boasts complete supporting facilities in terms of commerce, education and leisure, and is a top hillside mansion of luxury and comfort.

Landscape design follows the Spanish style of the building. The ebullient color of ground coverings, dynamic water spray landscape, swaying flowers and natural bushy palm trees interplay with each other, working together to create a Spanish style landscape. Pergola, end post, flowerpot and other details are well designed to create a unique romantic fashion. In addition, artificial works are integrated into natural beauty to construct sunny, ecological, romantic Spanish style residential community.

悠兰山位于惠博沿江生态走廊，高踞于城市新中心，将厚重优雅的欧洲山镇建筑风格与中国传统居住哲学融为一体，创造出独特的山地人居聚落。建筑在半山向阳坡地，占据城市制高点，俯瞰城市，远望东江，10 分钟双城圈，珠三角 1 小时生活圈，商业教育休闲配套一应俱全，是集豪华舒适于一身的顶级半山豪宅。

景观设计延续建筑的西班牙风情，热情奔放的地面铺装颜色，热力动感的喷水景观，摇曳生姿的花朵与自然而浓密的棕榈树种结合在一起，相映成趣，共同构筑了西班牙式的风情景观。通过对廊架、端柱、花钵等细节的精心设计打造出浪漫时尚的独到品味，融人工斧凿于自然美景中，营造阳光、生态、浪漫的西班牙风情小镇生活社区。

1
2

1 exquisitely designed landscape nodes
2 entrance of the building

1 精致大气的节点景观
2 建筑入户处理

1	3
2	

1 zigzag garden path
2 comfortable and pleasant recreation space
3 waterfront landscapes bring people close to water

1 充满意境的蜿蜒园路
2 舒适怡人的休憩空间
3 亲水设计提高景观参与性

1 landscape swimming pool
2 delicate soft decorations
3 pay attention to creating architectural and landscape spaces

1 风情大气的景观泳池
2 精致的软装设计
3 强调建筑与园林空间感的营造

1		4	5
2	3	6	7

1 materials and details
2 spanish style green landscape
3 skillful treatment of the altitude difference
4 landscape at the building entrance
5 paved square
6 sculptural lights
7 lawn for relaxing and playing

1 材料搭配和细节刻画
2 绿化强调西班牙风格的热情
3 大气仪式感的高差处理
4 入户景观元素丰富
5 风情精致的铺装广场
6 雕塑化灯具提升景观效果
7 参与性草坪空间

1		4	5
2	3		6

1 half indoor space shows high quality of the project
2 landscape of the elevated floor
3 corridor of the elevated floor
4 well-proportioned cascade
5 art-deco design can be found everywhere
6 simple and elegant water features

1 半室内空间尽显品质感
2 架空层景观
3 架空廊道处理形式
4 尺度怡人的跌水景观
5 无处不在的装饰主义设计
6 简洁大气的对景水景

Yunda Central Plaza Residential Area, Changsha
运达中央广场居住区，长沙
——寄情自然山水，独享武广生态大宅；崇尚典雅奢华，共拥顶级财富商圈

Developer: Hu'nan Yunda Real Estate Development Co.,Ltd.
Project Type: Integrated Residential Area
Project Area: 92,280 m^2
Design Content: Landscape Conceptual Design & Production Design
Design Period: 2010

Yunda Central Plaza is located at the heart of Wuguang business circle, Changsha Road, Changsha, near the subway entrance, and has very convenient traffic. It is composed of high-end commercial buildings and high-grade refined decorated houses, including two super luxurious five-star hotels of W Hotel and Regis Hotel, large international famous brand stores, 5A international office buildings and international hotel apartment. The introduction of the international fashion design concept of "low carbon" bring success for the future Changsha low carbon model building and fashion life of the ecological community.

There are light green buds in spring, rainy days in summer, falling leaves in autumn and thick snow in winter. Flowers of spring, rains of summer, fallen leaves in autumn, white snow in winter. We don't need to experience the life with murmuring stream and singing birds in wilderness; the real high-quality living environment will enable us to find the inner peace even in the busy city. Upon deep consideration, we decided to introduce the stream water into the garden. Flowing water makes the whole garden full of energy and vigor. The community thus becomes quiet and comfortable with magic changes of four seasons.

委托单位：湖南运达房地产开发有限公司
项目类型：综合居住区
项目面积：92 280 m^2
设计内容：景观方案及施工图设计
设计时间：2010 年

运达中央广场位于长沙市长沙大道武广商圈核心地段，地铁出入口附近，交通便利。运达中央广场由高端商业和高档精装修住宅组成，包括 W 酒店和瑞吉酒店两大超豪华五星级酒店、大型国际名品商场、国际 5A 写字楼以及国际酒店公寓。设计引入了国际流行的“低碳”理念，成就未来长沙低碳建筑样板和生态社区典范的时尚生活高地。

春天繁枝容易纷纷落，夏天连雨不知春去，秋天梱庭多落叶，冬天深夜知雪重。倾听潺潺流水、鸟鸣虫叫的生活，不一定要到林泉野径去才能体会得到，更高层次的生活环境，应该是在都市繁华之中寻找心灵的净土。引溪入园，是我们深思熟虑后的一个设计理念，流动的溪水，让整个园林活跃起来，变得更加有生命力，让小区显得更加的宁静安逸，同时也让小区四季变化充满魅力。

1-2 main entrance of Yunda Central Plaza

1–2 运达中央广场主入口实景

1 | 2 | 3
  | 4 | 5

1 overlooking the waterfront wooden platform
2 waterfront wooden platform in the night
3 corridor connecting all buildings
4 wooden platform
5 zigzag plank road

1 俯瞰临水休闲木平台
2 休闲亲水木平台夜景
3 户户连通，风雨连廊
4 休闲亲水木台
5 曲折变化的自然木栈道

1	
2	4
3	

1 landscape pavilion and cascades
2 "Water Curtain Cave"
3 night views of the residential community
4 lively music fountains

1 层层的跌水水景烘托出视线中心的景观亭
2 “水帘洞”乘凉空间
3 小区夜景实景
4 音乐喷泉活跃了全园的静谧氛围

1	2	4	
3		5	6

1 copper crane sculpture
2 copper twin-fish sculpture
3 landscape of the backyard meeting area in the sales center
4 overlooking the garden
5 sunshine lawn
6 waterfront wooden platform

1 仙鹤铜雕
2 双鱼戏水铜雕
3 售楼部后场洽谈区景观
4 居高俯览园区景观
5 阳光大草坪
6 亲水木平台

Zhujiang Sun Town, Chongqing
珠江太阳城，重庆
——山水清幽处，写意都市间

Developer: Chongqing Zhujiang Industrial Co., Ltd.
Project Type: Integrated Residential Area
Project Area: About 280,000 m^2
Design Content: Landscape Conceptual Design & Production Design
Design Period: 2006

委托单位：重庆珠江实业有限公司
项目类型：综合居住区
项目面积：约 280 000 m^2
设计内容：景观方案及施工图设计
设计时间：2006 年

Zhujiang Sun Town is located in Liujiatai section, Binjiang Road, just at the golden section point of the CBD core area and the riverbank ecological hinterland; it faces Jialing River in front, backs to the Wuli Store, closes to Huanghuayuan Bridge and looks over Yuzhong Peninsula in the distance; the overall planning style of the project is positioned as leisure living community of Bali type. Landscape design, combining with the local climate characteristics, makes full use of slope terrain for landscaping to minimize the damage to the original topography. According to the people's nature of living near by water, it takes water system as the main axis and plant landscape as the fundamental key to form mountain landscape, which successfully introduces waterside culture into the mountain beautiful scenery and becomes a new model for the enjoyable life in ecological city. The buildings hide naturally among the mountains and water, river and the sky set each other off, forest of green trees and gentle breeze blows, all bring the busy city life with natural, pure and fresh air.

珠江太阳城位于江北滨江路刘家台路段，身处 CBD 核心区域与江岸生态腹地的黄金分割点，前临嘉陵江，背靠五里店，紧邻黄花园大桥，与渝中半岛凭栏而望，项目规划整体风格定位为巴厘岛式休闲度假风情的居住社区。

景观设计结合当地气候特征，充分利用坡地地形造景，尽量减少对原地形的破坏。利用人们择水而居的天性，形成以水系为主轴、以植物造景为基调的山地园林景观，将亲水文化引入山城美景中，成为生态化城市写意生活的新典范。在这里，建筑自然隐匿于山水之间，江水天色相映生辉，绿树成林，柔风拂面，为繁忙的都市生活凭添了自然清新的空气。

1
2
3

1 well arranged tall and short plants
2–3 section of swimming pool

1 造型各异，高低起伏的植物群落
2–3 泳池剖面图

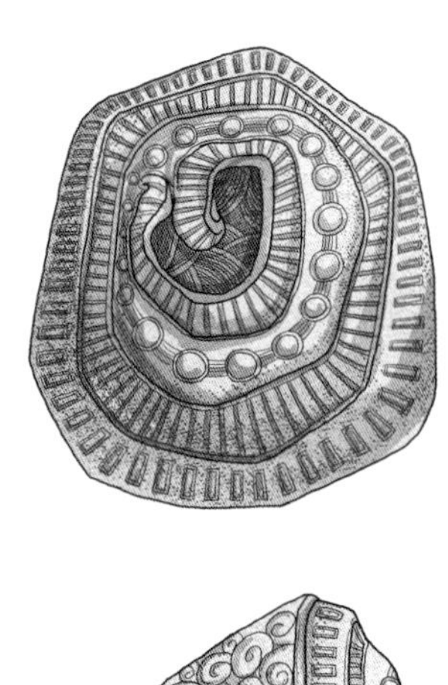

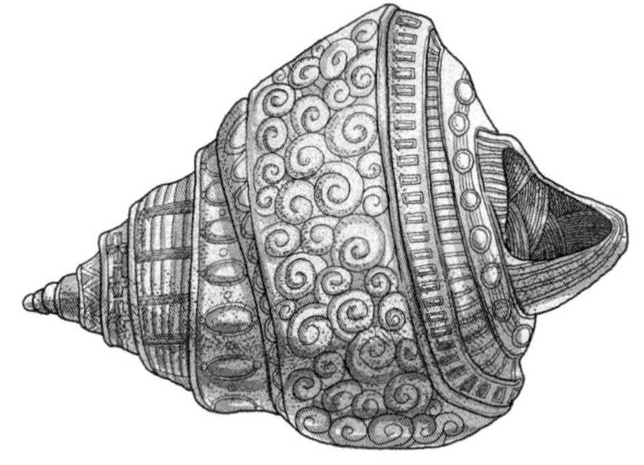

珠江太阳城，重庆

1	2	3
	4	5

1 sculptures in water
2 hand drawing of conch sculpture
3 hand drawing of the beautiful peacock sculpture
4 leisure platform, stepping stones, sculptures and landscape wall around the swimming pool ...
5 view from one edge of the swimming pool

1 水中阵列雕塑
2 海螺雕塑手稿
3 尾羽华丽的孔雀雕塑手稿
4 泳池区丰富的景观：水中休闲平台、汀步、雕塑、景墙……
5 泳池一端视角

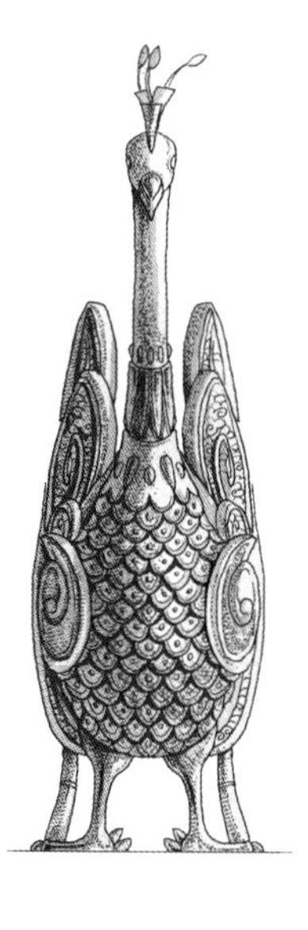

Sincere Star Metropolis, Chongqing
协信星都会 · 溪山墅，重庆
——溪山掩映斜阳里，墅台影动景致深

Developer: Sincere Group
Project Type: Integrated Residential Area
Project Area: 131,238 m^2
Design Content: Landscape Conceptual Design & Production Design
Design Period: 2011

委托单位：协信地产集团
项目类型：综合居住区
项目面积：131 238 m^2
设计内容：景观方案及施工图设计
设计时间：2011 年

The project is located in Zhaomu Mountain of Chongqing, close to Zhaomu Mountain Park and Sports Park, enjoying beautiful landscapes and quiet environment. The design has taken advantage of the big height difference within the site, integrated eco environment with modern design, and created an ecological community for healthcare and vocation.

Following the Art-Deco architectural style, the landscape designers have created a series of terraced green spaces according to the topography. It has realized the harmony between technology and human care.

Green Park: the wide green belt around the community forms two barriers to insulate sound and dust. In addition to the micro topography, the varied green spaces and the zigzag footpaths combine to create a green park which provide more green spaces for activities.

Clubs and Villas Area: As the meeting point of the community, the club is designed with resort-style landscape spaces. Sitting on the terrace of the club, one can enjoy the beautiful swimming pool and valley beneath it. The leisure platform beside the swimming pool allows people to enjoy the beautiful views. Flowing water and mist spray creates a quiet atmosphere. The private plank road leading to the valley will bring people close to nature.

High-rise Area: traditional Chinese landscape skills are employed to create a series of green spaces for the high-rise area. Sunken spaces and slopes are created to provide more landscape spaces. Zigzag footpaths will lead people to experience the park life in the community.

项目位于重庆照母山片区，临近照母山公园、体育公园等，自然环境优美，是繁华之处难得的清静之地。区内高差较大，设计因地制宜，化劣势为优势，突出生态境地和现代感的融合，打造养生度假生态社区。

景观设计师提取 Art-Deco 线条型的建筑立面装饰风格元素，并结合场地的地形地貌，以自然层级式绿化为风格基调，结合现代主义理性的方式，以舒适典雅韵味来探寻技术美与人性关怀的和谐统一。

绿带公园：社区外部围绕着宽阔的绿地，形成两个隔音隔尘屏障，设计中通过地形的塑造、绿化空间的变换和散步径的穿插，打造舒适的绿带公园，为居者提供更多绿色活动场地。

会所与别墅区：会所作为社区人流的枢纽，利用高差形成了度假酒店式的观景空间，在会所的露台上可以俯瞰泳池和溪谷的幽深景色。泳池边的休闲平台提供了令人陶醉的观赏空间。以水声和喷雾营造远离尘嚣的宁静氛围，同时可通过私家栈道下到溪谷，享受流动的清波，回归自然的养生生活。

高层区：设计中运用了障景、借景、框景等手法设计绿化空间，利用部分下沉、堆坡结合绿化的方法，在有限的空间中创造无限的景观空间。曲径通幽，在步移景异中体会公园生活的写意。

open forest and lawn as well as blooming flowers　开阔的疏林草坪和盛放的鲜花

1	3
2	4

1 welcoming fountains
2 main entrance of the sales center
3 garden path full of flowers
4 pool and fountain in front of the sales center

1 涌泉营造热烈的迎宾氛围
2 售楼部主入口
3 鲜花簇拥的园路
4 售楼部前涌泉水池

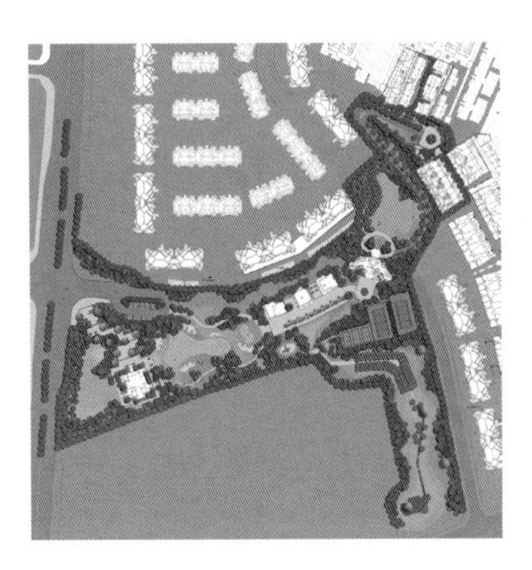

Jinke City, Chongqing
金科城，重庆
—— 托斯卡纳阳光小镇

Developer: Jinke Real Estate Group
Project Type: Low-density Residential Area
Project Area: 159,620 m^2
Design Content: Landscape Conceptual Design & Production Design
Design Period: 2013

委托单位：金科地产集团
项目类型：低密度居住区
项目面积：159 620 m^2
设计内容：景观方案及施工图设计
设计时间：2013 年

Located at the central Liangjiang New Area, Jinke City is near the Jialing River, Jiayue Bridge and Caijia Railway Bridge, overlooks Yuelai International Convention Center, Yuelai New Town on the other side of the river and borders Lijia Commercial District to the south. One may get to Chongqing bonded port by heading northward for 15 minutes. It boasts advantages of Caijia, Lijia and Yuelai Areas, connects multiple city main roads and annexes surrounding Guanyinyan Park, Caijia Central Forest Plaza, Sports Park…forming a breathtaking urban forest landscape.

Following the design concept "flowing landscape—poetic landscape space", designers start the work in line with local conditions. Open space is created with lawns and big trees in the wide area, while semi-open space is accompanied with waterscape, viewing platform and footpath under the tree in the narrow area. On the way to the hillside villa, landscape wall and rails are designed to break the cramped sense. Impressive lawns & sea of flowers and colorful romantic boulevards provide a full sensory and multi-level landscape experience, showing the cultural atmosphere of an international community and creating a fresh and natural scene of life.

金科城位于两江新区核心，嘉陵江畔、嘉悦大桥旁，蔡家轨道大桥附近，对岸即是悦来国际会展中心、悦来新城，南望礼嘉高尚商业区组团，往北 15 分钟可达重庆保税港区，汇聚蔡家、礼嘉、悦来三大区域的优势，为多个主城区域连接要道，周边观音岩公园、蔡家中央森林广场、体育公园……形成叹为观止的城市森林生态景观。

项目遵循"流淌的景观——诗一样的景观空间"的设计理念，因地制宜，在宽阔区域以草坪、大树等形成开敞空间，狭小的区域以水景、观景台、林下休闲步道形成半开敞空间，在通往坡地别墅的上山步道上设计景墙、栏杆等打破整个空间的局促感。大气震撼的草坪花海，浪漫缤纷的林荫道，形成全感官、多层次的景观体验，展现出大气的国际化社区文化，营造一种清新自然的生活场景。

1 fairytale wedding ceremony
2 funny plank road
3 colorful views along plank road

1 童话般的婚礼场景
2 野趣的景观木栈道
3 丰富多样的栈道场景

Bread Today
Bread Today

1 | 2 | 3 / 4

1 commercial space and church for wedding
2 clever use of small spaces and altitude difference
3 colorful villa courtyard
4 pergola and soft decorations

1 商业空间与婚礼教堂
2 小空间与高差的巧妙配合
3 生活色彩的别墅院落
4 景观廊架配合丰富软装设计

1 pleasant and funny space
2 nursery garden and plank road

1 心旷神怡的野趣空间
2 苗场与景观栈道的借景配合设计

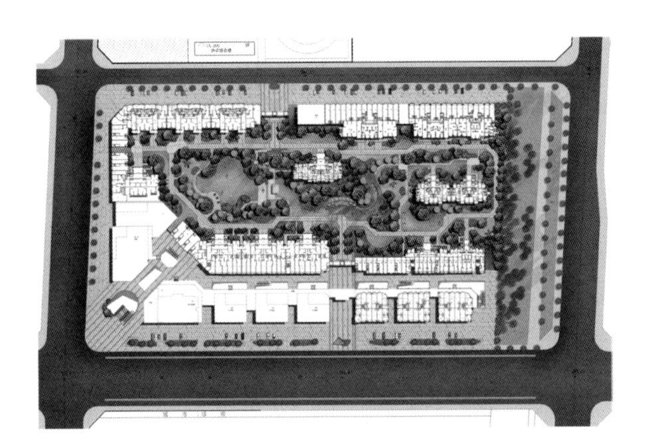

Jinke Glamour Community, Xi'an
金科·天籁城，西安
——城市中的一则天籁

Developer: Jinke Real Estate Group
Project Type: Integrated Residential Area
Project Area: 62,128 m^2
Design Content: Landscape Conceptual Design & Production Design
Design Period: 2013

委托单位：金科地产集团
项目类型：综合居住区
项目面积：62 128 m^2
设计内容：景观方案及施工图设计
设计时间：2013 年

Built on the site of Daming Palace, the Glamour City is the first Jinke project in Xi'an city. Modern design skills are employed to integrate modern art elements with natural spaces. It is going to build a happy homeland for Xi'an citizens with rich space experience, exquisite landscape features as well as local culture.

Simple and elegant broken-line elements are used to create open and close spaces. Designed with the theme of "modern elegant life in rapidly developing city", it well arranges every detail to achieve the harmony between buildings and landscape.

It tries to create deep impressive spaces with landscape groups, large pools and large lawns. Art-deco architectures, pavements and art decorations highlight the quality of the project and increase fun to the spaces. All the landscape spaces are well designed with exquisite details to meet different functional requirements.

金科·天籁城位于西安大明宫遗址，是金科在西安的第一个项目。设计中运用现代简约的设计手法打造出现代风格的景观，力求将现代艺术而精致的元素和疏密有致的自然空间相互融合。通过丰富的空间体验、精致的小品，结合西安的人文特色与历史底蕴，打造适宜西安人居住的幸福家园。

景观设计采用简洁明快的折线形式，呈现空间的开、闭、合等形态，在设计上充分考虑近景、远景等多层次的空间视线组织，以此达到最佳观感。整个小区景观围绕“现代简约生活，都市明快主张”的主题进行规划设计。点线面、不规则几何图形的有机结合，使得整体空间层次明确，风格统一，使建筑与景观达到了完美的相融。

设计中利用组团园林空间，尽可能营造有纵深感的大空间，以大水面、大草坪给人以震撼力，同时突出 Art-Deco 的艺术气息，通过构筑物、铺装等精致处理，突出项目的品味，并配合艺术小品的点缀，丰富空间的趣味性。在营造大空间的同时做到疏密有致，疏朗中见细节点缀，密林中见幽深恬静，满足各种活动的功能需求。

1
2 | 3

1 simple and elegant landscape wall with logo
2 broken line-shaped bench
3 landscape details with broken lines

1 异形 logo 景墙造型简洁明快
2 折线坐凳充满景观趣味
3 景观细节充满折线形式

1 special-shaped benches for sitting and communication
2 sunshine lawn
3 long and narrow boulevard
4 outdoor bar counter
5 outdoor table and seats

1 异形坐凳提供新交往空间
2 阳光草坪
3 林荫夹道纵深感极强
4 户外吧台洋溢生活热情
5 户外时尚卡座

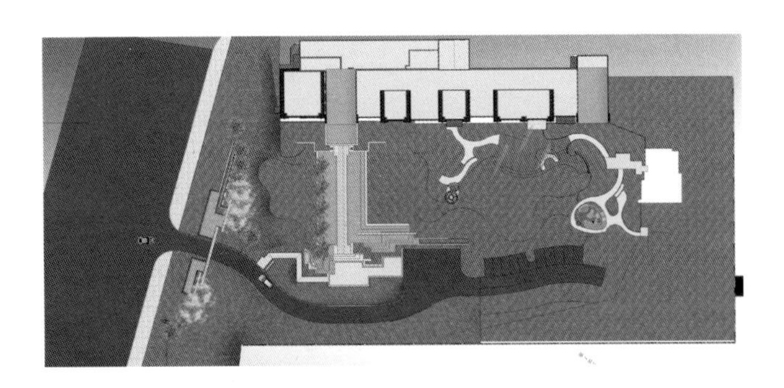

Longfor Chianti International, Xi'an
龙湖 · 香醍国际，西安
——在风景中升华自己，故事延续

Developer: Longfor Properties Co., Ltd.
Project Type: Integrated Residential Area
Project Area: 11,704.77 m^2
Design Content: Landscape Conceptual Design & Production Design
Design Period: 2015

委托单位：龙湖地产有限公司
项目类型：综合居住区
项目面积：11 704.77 m^2
设计内容：景观方案及施工图设计
设计时间：2015 年

Xi'an Chianti International is located in Chanba Eco Park, boasting a total residential area of 1,500,000 m^2. The exhibition area is designed with Longfor's typical five-level landscape system. The front yard, which opens like an urban living room, is decorated by waterscapes on two sides. Hard landscapes and soft plants, open and closed spaces, different colors, vegetation and lawn, all combine together to present a beautiful and elegant image.

The landscape design tries to get the balance between smooth and beauty as well as simplicity and elegance. Exquisitely designed landscape items can be found every where. All activity spaces are well organized to provide different experiences and remind people of beautiful memories.

Following the architectural style, the landscape is designed in modern luxurious French style. The front yard is symmetrical along the central axis, and the entrance is characterized by grass-covered steps and landscape walls of different heights. A zigzag garden path leads people to the front yard, along which, large area waterscapes well decorate the way to the sales center. The back yard spaces are well connected by garden path, wooden platform and children's playground.

西安龙湖 · 香醍国际位于浐灞生态园中，拥有 150 万平方米国际滨河住区，示范区注入龙湖独有的五重景观体系。前场以都市会客厅的门户形象拉开帷幕，两侧迎宾水景营造轻松悦动氛围，合理的硬景及精神种植、疏密的开合空间及融洽的色彩过度，植被与干净大草坪，相互呼应，彼此衬托，唯美而雅致。

整个展示区园林在营造过程中追求平滑与柔美的协调，飘逸与质朴的对比。大气的尺度感下，处处都有细致入微的小景致。多重体验空间有序组织、收放自如，将情怀融入景观，艺术、自然融合于活动空间，唤醒人们内心的记忆。

建筑设计以现代法式为基调，景观设计与建筑相呼应，打造气势恢宏，豪华舒适的贵族风格。展示区前场布局上突出轴线的对称，展示区入口利用层级草阶和高低错落的景墙强调入口的展示面，进入展示区前场要通过一条曲径通幽的林荫路，两边大面积的水景将进入售楼部的道路悬浮在水中；后场主要以自然流畅的园路、木平台以及丰富的儿童活动区贯穿起来。

1
2

1 welcoming waterscape axis at the sales center
2 spacious space with waterscape and lawn

1 销售中心迎宾水景轴线
2 宽阔的水景草坪空间

香醍国际

龙湖·香醍国际

前方学校
减速慢行

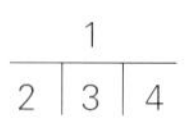

1
2 | 3 | 4

1 magnificent landscape interface at the entrance
2 French-style landscape wall and hallway in the front square
3 flowers, waterscape and landscape wall echo each other
4 characteristic logo

1 恢弘大气的入口景观界面
2 法式风情的前场景墙门厅
3 时花、水景与景墙相互呼应
4 特色 logo 标识

1	4
2 3	5

1 children's activity space
2-3 distant view of the children's playground
4 lighting effect on the landscape walls
5 the sunlight plays upon the surface of the water

1 结合地形营造儿童活动空间的趣味感
2-3 儿童活动场地远景
4 景墙灯光效果
5 阳光投射到层级水面，更显优雅风情

龙湖·香醍国际，西安

1	2	5
3	4	6

1-4 details of waterscapes
5 wooden platform under the light
6 seats and tables with broad views

1-4 不同场景下的水景细部同样动人
5 灯光掩映下的后场木平台尽显静谧氛围
6 视野广阔的休闲卡座

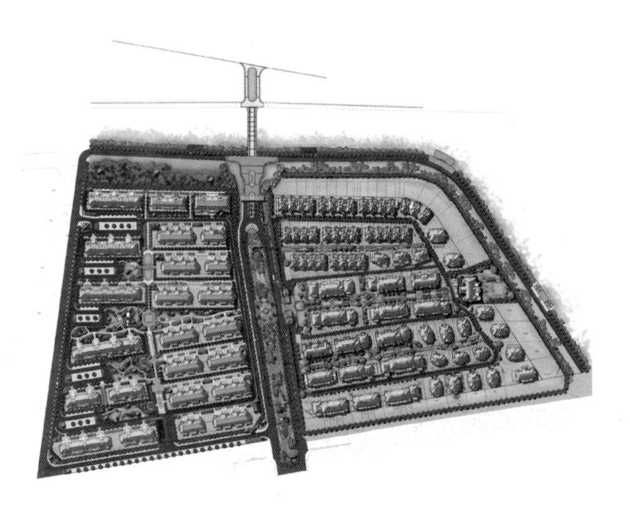

Jinke Royal Spring Villa, Beijing
金科王府，北京
—— 收藏法式庄园，礼赞秩序人生

Developer: Jinke Real Estate Group
Project Type: Low-density Residential Area
Project Area: 158,000 m^2
Design Content: Landscape Conceptual Design & Production Design
Design Period: 2011

委托单位：金科地产集团
项目类型：低密度居住区
项目面积：158 000 m^2
设计内容：景观方案及施工图设计
设计时间：2011 年

The project was titled "2013 China International Quality Apartments"; "2012 the Benchmark Villa Project of Beijing"; "2012 China's Top 10 Luxury Residence"

该项目荣获"2013 中国国际化品质楼盘"；"2012 北京房地产行业标杆别墅"；"2012 中国十大最具价值豪宅"奖。

Royal Spring Villa is located in Changping District of Beijing, a core position in the traditional Xiaotangshan hot spring resort; it is close to the North Sixth Ring in the south side, only 1 km away from the Patio project in the west, 3 km away to the Beijing-Chengde Highway in the east, which is a rare low-density high-end residential community in Aobei Area. The community is mainly set with 154 detached, semi-detached villas and townhouse, and matched with luxury viewing houses. The villas take French style as the architectural form with elaborate community royal garden, forming strong French amorous feelings. The project faces Hulu River, the affluent of Wenyu River on both sides of the east and north, and bears unparalleled wetland condition and more than one hundreds over-100-year-old trees, making the entire project become a veritable "natural oxygen bar". Around the project there are mature residential communities and high-grade leisure facilities, such as Napa Valley, LongMai Resort & Spa, Xiaotangshan Sanatorium and Jiuhua Resort & Convention Center, etc.

With great respect to nature and ecology, Jinke Wangfu also pays attention to the comfort of living. Lawns, flowers, bridges and pavilions are skillfully designed to create a picturesque ecological park which allows the residents to enjoy the prosperity of the city and the tranquility of natural life.

Royal Spring Villa has integrated the landscape skills of Chinese traditional royal garden and the palace of King Louis XIV to create a harmonious green palace. The overall layout is symmetrical with emphasis on the relationship between road landscape and group landscape. Diversified green trees, carved flowerpots, tall and short plants combine together to create a series of dynamic landscape spaces. The garden is not only a place for strolling but also a high-class venue for meeting and banquet.

1 French-style villa
2 details of the flowerpot

1 法式别墅，风情浓郁
2 花钵细节

金科王府位于北京市昌平区，传统的小汤山温泉度假区核心位置；南侧紧邻北六环，西距帕提欧项目仅 1 公里之遥，东距京承高速 3 公里，是奥北地区不可多得的低密度高档住宅社区。小区以 154 套独栋、双拼和联排别墅为主，并配以豪华观景洋房。别墅的建筑形式为法式风格，配以精心雕琢的社区宫廷园林，从而形成浓郁的法兰西风情。项目东、北两侧濒临温榆河的支流葫芦河，湿地条件不可复制，更兼之小区内部百余株百年以上树龄的参天古树，使得整个项目成为一个名副其实的“天然氧吧”。在项目周边，遍布成熟居住社区和高档休闲设施，如纳帕溪谷、龙脉温泉度假中心、小汤山疗养院、九华山庄等。

金科王府在充分尊重自然、生态的基础上，不忘打造人居的舒适性。配合万平原生密林，精心规划草坪、花卉、栈桥、亭台，打造原生态漫步公园，婆娑树影、横卧栈桥、亭台楼榭交相辉映，一副天然无忧的长生画卷。出，则繁华，把控事业大局；入，则静谧，独有的领域和天地，让主人享受王者归来的生活佳境，以和谐、尊崇的姿态拥抱自然生活。

1	3	4
2	5	6

1 french classical green carpet
2 flowering trees
3-6 outdoor leisure space

1 法式园林中的经典绿毯
2 热情奔放的观花乔木
3-6 户外休闲空间

金科王府，汲取勒诺特式宫廷庭园之精髓，将传统中式皇家园林与路易十四宫廷庭园手笔相融合，打造与王府主体的宫殿式建筑完全协调的一座“绿宫”。整体布局强调中轴对称与均衡，重视律式的严谨性，重视道路景观与每个组团区域绿地的关系，并结合绿化树种的多样性，宫廷模纹花坛变化万千，与高差层次的植被编织出步移景异的视觉空间，给庭园真正雕刻出留念性建筑物的特征。园林，已经超出了一般意义上的漫步场所，也为高级会见和盛大宴会而生。

1	2	5
3	4	

1 elevation of the villa
2 soft flowers and hard buildings
3 forest land covered by flowers
4 experience French style under tree
5 colorful plants beside villas

1 别墅立面
2 鲜花与建筑的刚柔对比
3 林下地被缀满花朵
4 树荫下，品味法式时光
5 别墅旁植物造景注重引入丰富的色彩

Dongsheng Dawn Garden, Zibo
东升 · 曦园，淄博
——院落三进，清流几许

Developer: Dongsheng Real Estate Development Co.,Ltd.
Project Type: Low-density Residential Area
Project Area: 83,136 m²
Design Content: Landscape Conceptual Design & Production Design
Design Period: 2013

委托单位：东升房地产开发有限公司
项目类型：低密度居住区
项目面积：83 136 m²
设计内容：景观方案及施工图设计
设计时间：2013 年

Zibo is a city in northern China with a long history and four distinct seasons. It an open city in the development of modern industry. Just like the city, people living here are open, magnanimous and innovative.

Sunlight Garden is the epitome of the city.

New Orientalism means the renaissance of oriental culture, which presents the oriental aesthetics, modern western spirit, the designer's thinking and the characteristics of the times. Skillful designers will integrate traditional Chinese culture and elements into modern living spaces in an appropriate way and make them the soul of the work.

Taking advantage of the site conditions and inspired by the oriental living ideas of "front and back yard, park residence", the designers have created a new landscape environment integrating traditional residence and modern park-style community. Classical garden skills are employed to create beautiful courtyard spaces with modern landscape walls, water features, pavilions and so on.

「Courtyard-style Entrance, Aristocratic Residence」

Sunlight Garden is inspired by traditional Chinese residence to well organize the enclosed courtyard space for privacy and the open spaces neighboring communications. The unique courtyard space of Sunlight Garden perfectly interprets the elegant spirit of traditional oriental residence.

The entrance of the Sunlight Garden is simple and elegant with a landscape wall, a cascade and two symmetrically planted trees. The front yard of Sunlight is designed with reflection water and Chinese traditional water features. In the center of the front yard, a welcoming pine tree stands quietly to tell the story of the courtyard.

Walking along the corridor from the front yard to the central yard, there is a cascade and a Chinese-style pavilion. Elegant lamps will lead you to the reception hall and the well-organized spaces will enable you to experience the distinguished quality of hotel. Pass through the hall you will step into the open

1 lights and shadows
2 courtyard of the club

1 灯影交辉的夜色
2 会所庭院的景深

backyard where big trees stand in row by the swimming pool and pergola dialogues with the blue sky.

「 **Landscape Changes Step by Step** 」

Walking along the zigzag garden path, landscape spaces will change as you move on.

「 **Stream and Bridge** 」

A gurgling stream runs from east to west to form natural revetment. Artificial hills, exquisite vegetation, gurgling stream, quiet stone bridge, water plants and green trees present a poetic image of North China's water village.

1	4
2 | 3	5

1 courtyard of the club
2 lively cascade and the pergolas over the reflection water
3 hotel-style brazier landscape at the entrance
4 cascades at the entrance
5 cascades and pergolas

1 会所庭院
2 生动的跌水与坐落在镜面水上的休息廊架相呼应
3 以极具酒店风情的火盆元素打造向心型入口景观
4 入口区跌水对景
5 瀑布跌水景观廊架

淄博的四季是分明的，这座北方的城有着悠长的历史积淀，又是现代工业化崛起的经济开放城市。生活在这里的人们具有历史人文主义情怀，同时又开放包容，具有创新精神。

曦园，正是这座城市性格的浓缩和体现，传承与创新。人们说，曦园，是一场人居革命。

“新东方主义”代表是一种东方文化本底精神的回归，是一种东方文化的复兴运动。其作品往往涵盖了东方的审美传统，西方的现代精神，设计师的思考创造，时代的浓烈色彩。它是设计师把中国传统文化精髓与元素“揉”进适合现代生活方式的空间当中，其力道与手法体现着设计师的功力，也是整个作品的灵魂所在。

设计师利用项目优越的规划条件，融入东方传统人居观念，以“前庭后院，家傍公园”为设计意念，打造出传统居庭与现代公园式社区相结合的居住新景观。运用借景、对景、框景、障景等古典造园手法，营造出开合有致的深深院落，园林景观方寸内尽现乾坤。现代的景墙、水景、景亭等元素镶嵌点缀其中，一曲低吟浅唱双重合奏，全新演绎“庭院深深深几许”，

“闲花深院听啼莺”的园林意境。

『院落式入口，大宅风范』

中式传统居庭讲究空间礼序和邻里组群关系，曦园正是汲取了传统宅院布局精髓，把围合庭院与开放空间有序组织，并与现代住区空间的共享属性以及私属领域的关怀需求巧妙结合。曦园的成功在于抓住了东方居庭独特的院落式空间“前庭后院”、欲扬先抑的特点，颇有循序渐进的儒雅精神。

曦园的门堂大气简约，细节中不失中式韵味。景墙跌水与两株对植大树，打造出尊贵而内敛的气质。穿过门堂，一方宁静水厅印入眼帘，这便是曦园的前庭。既有镜面水的空灵纯净，又有中式传统水庭的幽幽禅意，正是两者结合创新的产物。水厅中央静立一株迎客松，似乎正与纷纷到来的宾客娓娓诉说着庭院里的故事。

沿连廊闲步前行，过水厅至中庭，在面前展开的跌级水景，矗立着一轻盈而又饱含中式韵味的休闲亭。而亭边树影婆娑，一抹闲情逸致油然而生，精致而简洁的灯饰，

把人的视线引向接待大厅，层层递进的空间处处体现酒店般尊贵的品质。穿庭过厅，步入后院，豁然开朗。沿泳池边阵列排布的大树形成夹景，将视线引向远处的天空，同时采用借后院景观廊架之景，形成对景，与古典造园手法一脉相承，妙不可言。

『叠石绿植，一步一景』

信步游曦园，园路婉转，地形起伏，草木葳蕤。景观空间开合有致、疏密相间。风景移步易景，千回百转，别有生趣。

『小桥流水，宁静致远』

一条潺潺的小溪沿着建筑展开，贯穿东西，自然蜿蜒的驳岸，错落的叠石与精巧的植栽勾勒出“小桥流水人家”的意境，近处汩汩流水声，远处的石板桥静卧水面，岸边的水草和绿树摇曳生姿，水面倒映出的光影画面，展现好一派北国江南水乡的诗情画意。

1	2	5
3	4	6

1 view of the cascade from the garden path
2 delicate paved garden path
3 garden landscape
4 zigzag garden path and the green spaces along it
5 overlooking the swimming pool
6 night view of the swimming pool

1 从园路窥视瀑布跌水景观
2 精致的铺装园路
3 地形丰富的园区景观
4 蜿蜒的园路，两侧绿化营造收放空间
5 从高处眺望具轴线感的泳池景观，更具震撼力
6 泳池夜景

Jinke Sunshine Town, Qingdao
金科·阳光美镇，青岛
——草原之所 生态栖居

Developer: Jinke Real Estate Group
Project Type: Large-Scale Integrated Residential Area
Project Area: 78,000 m^2
Design Content: Landscape Conceptual Design & Production Design
Design Period: 2014

委托单位：金科地产集团
项目类型：大型综合居住区
项目面积：78 000 m^2
设计内容：景观方案及施工图设计
设计时间：2014 年

Located in Qingdao Hi-tech Development Zone, the Sunshine Town is close to the agricultural park with an eco environment. The buildings are designed in Wright's prairie style, and the landscapes are designed accordingly to create poetic courtyards and paradises for wellness and happiness.

The exhibition area of Sunshine Town locates in the agricultural park, enjoying advantaged landscape resources. Sunshine, springs, trees, grass, flowers and the elaborately designed visiting route enable people to have immersive experience of the natural forest.

A successful design will adapt to the site and meet the requirements of the client's.

—— John Ormsbee Simonds

The exhibition area is designed with new ideas to interpret the uniqueness of the landscape. It creates a park-like environment for appreciation and experience.

The flower path leads to the elegant and rhythmic front square of the sales center where Wright's prairie style building sits on the landscape axis. The cascade dialogues with the landscape wall behind to highlight the luxury of the development. Hidden in the forest and lakes, the sales center stands in harmony with the surrounding environment.

Then we will step into the quiet inner square which is enclosed by exquisite landscape architectures and beautiful water features. Passing through the sales center, we will find the back yard dominated by the luxuriant white birch forest and green lawn lake.

The showflat area with flower path, green screens and eco spaces will provide great walking experience. At the entrance of this area, the geometrical landscape wall, the multi-level waterscape and the wooden path create a welcoming space. The designers follow the rule of nature and well organize the hills, waters, plants and landscape architectures to create a pleasant environment.

1 elegant sunshine lawn
2 leisure platform and fabric decorations

1 简洁精致的阳光草坪
2 休闲平台软装布品

Following the wooden steps, one will walk into the elegant showflat area and the open lawn. The lovely wedding lawn can also be used for outdoor banquet; the lounge in the east provides a private space for lovers. Flowers and shrubs, landscape and buildings, framework and wooden grid, all combine together to keep in harmony with the surrounding environment.

青岛金科·阳光美镇坐落于青岛高新技术开发区，近青岛示范农业园，生态环境良好。项目全区推崇建筑与自然和谐共生的理念，将草原式建筑特色与景观设计有机结合，并借景青岛示范农业园，以“生活有机的建筑、诗意生命的庭院、度假养生的乐土”为主题，构成一幅自然景观与建筑相互渗透的景观格调。

项目展示区位于农业示范园内，具有得天独厚的自然环境，农业园生态景观与自然精神完美融合。展示区景观表现形式自由而丰富，阳光、泉水、树林以及花草满布，精心规划的参观路线，人们在自然间轻松地享受着天地的恩泽，沉醉在这宛若原生态森林的世外桃源中。

成功的设计开始于因地制宜，以满足用户需求为终极目的。——约翰·西蒙兹

展示区以全新的设计思想来演绎住区景观的独创性与唯一性，利用项目地理优势形成穿梭公园回家的意境。不仅满足人的观赏性，更表达人对精神的更高层次追求。唯美的境界，心灵和环境融为一体。

经过花林夹道到达售楼部前广场，赖特式草原建筑坐落于景观轴线上，售楼部前广场简洁大气而又富有节奏感，外广场叠水景观与背后的景墙遥相呼应，营造高墙大院的豪宅气质。它掩映在森林、湖泊中，完美地与自然融合在一起。

步入台阶进入相对幽静的内广场，园建细部设计上形成统一精致的元素。采用现代围合形式，隔断外界的喧嚣与尘俗，通过四周环绕的水景给人以唯美静谧的氛围感。步入售楼部来到后广场，一片郁郁葱葱的白桦林仿佛滤去了城市的喧嚣，绿色流动的草坪湖空间作为视觉中心点。

样板房区，为步行体验区。花林夹道的小径营造出生态栖居的养生氛围，密植形成天然的屏障渲染浓厚生态感，立体流动的生态空间贯穿整个游览路线，让访客能在游览中感受若隐若现的美景。样板房区在入口处设置的几何背景墙与层级水景相结合，并与木质小径相连接以形成迎宾氛围。设计师师法自然，将山、水、植物与园建小品巧妙搭配，相得益彰。

层木质台阶欲扬先抑，使人在进入样板房区时顿感豁然开朗，开阔的草坪更是给人强烈的视觉震撼。甜蜜婚庆草坪可做婚礼户外用餐之用，而东侧酒廊架构则营造相对私密的甜蜜情侣空间。草、花、灌木渲染甜蜜氛围，构架设计与建筑元素相呼应，稳重的框架与木格相融合，与周边环境融为一体。

1	2	5
3	4	6

1 multi-level waterscape and wooden steps at the entrance of the backyard
2 european-style vegetation
3 zigzag garden path
4 well-organized platforms
5 warm wooden path
6 flower belts

1 层级水景结合木台阶后场入口
2 欧洲风情的植物配置
3 曲径通幽的园路
4 错落有致的平台
5 暖人的木园路
6 氛围浓郁的花带

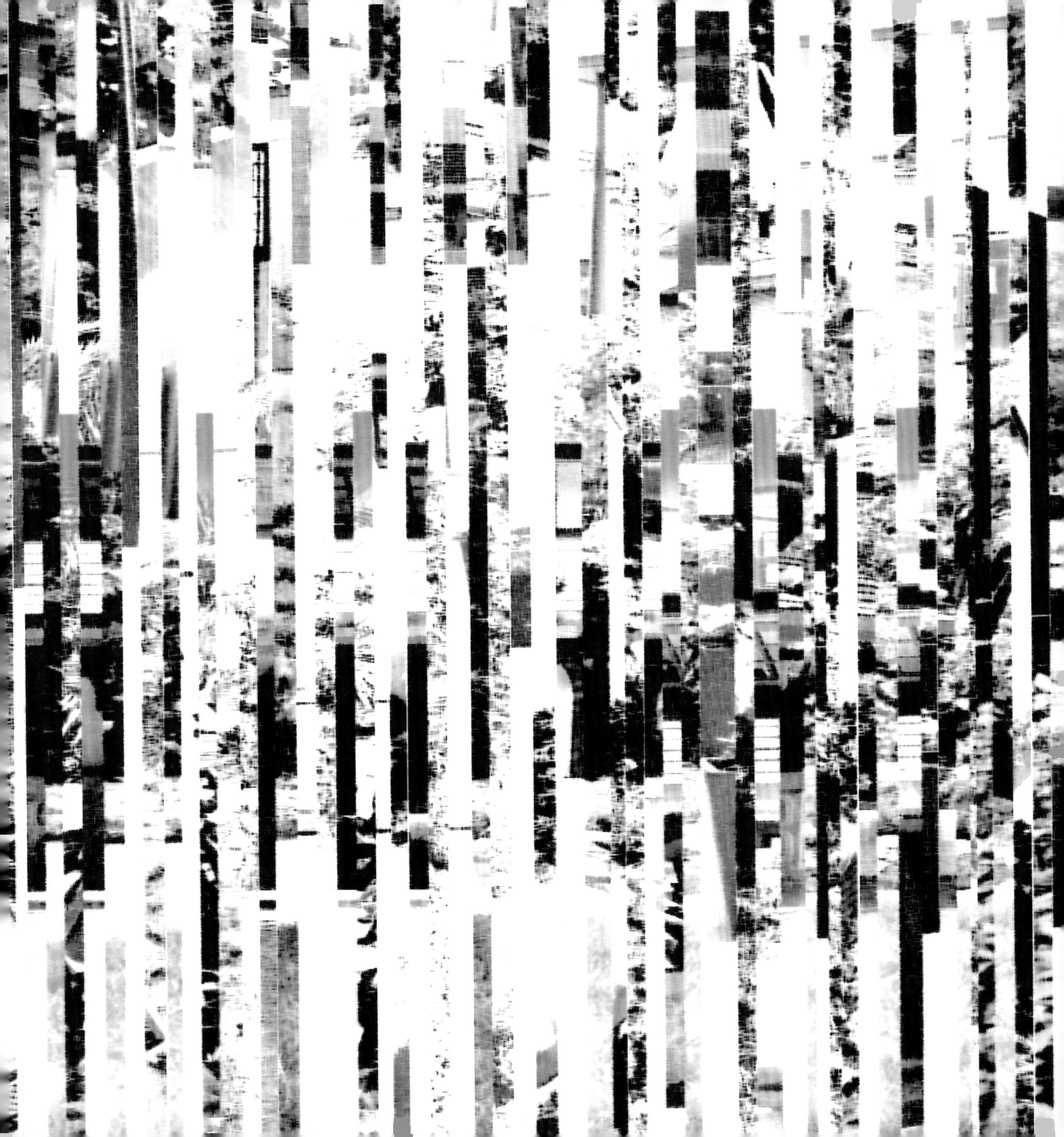

URBAN DEVELOPMENT

城市开放空间

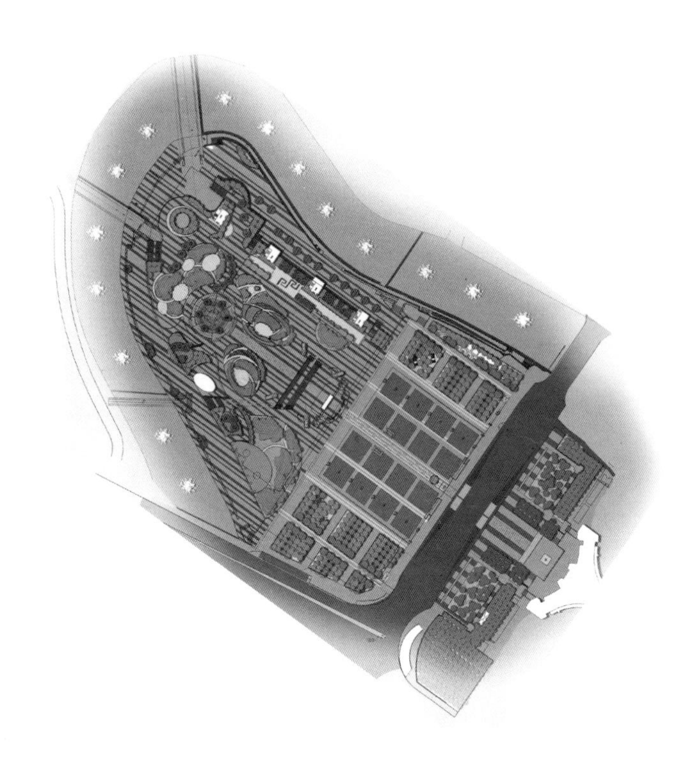

Zhucheng City Square, Guiyang
筑城广场，贵阳
——筑韵鸣钟，传颂盛世

Developer: Zhongtian Urban Development Group
Project Type: Urban Development
Project Area: About 87,000 m^2
Design Content: Landscape Conceptual Design & Production Design
Design Period: 2011
Architecture designed and planned by Teamer International
Main sculpture by Sculpture Design Company

委托单位：中天城投集团
项目类型：城市开放空间
项目面积：约 87 000 m^2
设计内容：景观方案及施工图设计
设计时间：2011 年
建筑及规划合作单位：天萌国际
主雕塑由雕塑公司提供

Located in the center of Nanming District, near to the north Ruijin South Road and to the west of Zunyi Road, the site of Zhucheng City Square is in the shape of a peninsula lying beside the beautiful Nanming River. With Guiyang No.1 Middle School on its north and Guiyang Civic Square on the south, Zhucheng City Square is connected with the surroundings by a bridge. The land is higher in the northwest and lower in the southeast with an altitude difference of 3~5m.

After soliciting opinions from the public, Guiyang government decides to build the "Zhucheng Square" here which will be the city's "living room" to promote the local culture and modern eco idea. It will be an ideal place for meeting, cultural activities, recreation and entertainment upon completion.

Guiyang is a multi-ethnic and cultural city with Han as the main nationality. In ancient times, Guiyang was rich in bamboos. The word "bamboo" in Chinese is pronounced as "Zhu", thus the city Guiyang was also called "Zhu" with a long history of bamboo culture. Therefore, the design of the central square embodies the history and culture of bamboo, uses a lot of local stones and plants, and creates some pavements, sculptures and bell towers to create an urban space which is modern, technological, ecological and sustainable.

Upon completion, the central square will be able to hold more than 50,000 people at the same time, therefore, reasonable circulation is essential. The solution is to build a vertical traffic system with two floors underground for parking. Moreover, new driveways will be designed to separate vehicles from pedestrians and ensure smooth traffic.

俯瞰位于南明区中心的项目基地，它呈半岛状依偎在风景秀丽的南明河畔，南临瑞金南路，东临遵义路，规划地块北部是原有的贵阳一中，南部为贵阳市人民广场，一座一中桥将其与周边相连。整体地形西北高、东南低，高差为3–5 米。

贵阳市委、市政府在广泛征求各方意见的基础上，拟在此建设“筑城广场”，将其打造成为贵阳的“都市客厅”，用以提升人文精神的象征性，展现当代生态观念，凝聚社会与自然活力。建成之后，这里将成为贵阳的政治集会和群众集会场所，同时也是市民们进行文化、休闲、娱乐活动的综合性空间。

贵阳，是一座以汉族为主的多民族聚居的城市，也是一座具有浓郁民族人文气息的城市。古代贵阳盛产竹子，“竹”

bird's-eye view of Zhucheng City Square 筑城广场鸟瞰图

与“筑”谐音，故贵阳简称“筑”，在长期的生产生活中积淀了深厚的竹文化。在中央生态广场的环境设计中，提取贵阳的历史、民族的竹文化因子，充分运用当地特色的石材和植物，通过铺装、雕塑、钟塔等景观元素，结合现代时尚的造景手法，将广场打造成为现代的、科技的、生态的、面向未来的城市空间。

建成后的中央生态广场能同时满足5万人以上的市民集会需要，因此交通组织也成为设计的难点与重点。设计师采取了竖向组织立体分流体系，设计了两层地下停车场，并构建新的车行通道，不仅让人车各行其道，也能让进入广场的车辆能迅速疏散。

1 | 2
1 | 3

1 night view of the square
2 colorful lights of the fountain pool
3 dazzling square in the night

1 筑城广场夜景俯视效果图
2 幻彩的喷水池灯光
3 夜幕下璀璨的筑城广场

1 | 4 | 5
2 | 3 | 6

1 towering "gold reed-pipe" sculpture
2 night view of Nanming River
3 night view of the square
4 landscape bridge runs across Nanming River
5 night view of Nanming River
6 lively auspicious animal sculpture

1 高耸入云的"金芦笙"
2 南明河夜景
3 原点广场夜景
4 横跨南明河的景观桥
5 南明河夜景
6 神采奕奕的祥兽雕塑

Poly Pazhou Eyes Greenland Park, Guangzhou

保利琶洲眼绿地公园，广州

—— 自然景观，水乡文化

Developer: Poly Real Estate Group
Project Type: Urban Development
Project Area: About 40,000 m^2
Design Content: Landscape Conceptual Design & Production Design
Design Period: 2012

委托单位：保利地产集团
项目类型：城市开放空间
项目面积：约 40 000 m^2
设计内容：景观方案及施工图设计
设计时间：2012 年

The park is ideally located in the center of Pazhou Island, on the south bank of the Pearl River. It is next to Guangzhou International Convention and Exhibition Center in the east, nearby the Pearl River on the north, and adjacent to a large empty land on the south. Upon completion, it will provide a great waterfront space for meeting, recreation and entertainment.

The landscape is designed in natural style with curved lines and simple materials. While the space design emphasizes people's participation and tries to provide more waterfront spaces. The plants are well selected to keep in harmony with the surrounding environment. In addition, the design of the park also interprets the local culture of Pazhou district and provides more cultural landscapes for those commercial areas.

保利琶洲眼绿地公园位于广州琶洲岛中部、珠江南岸。西面为广州国际会展中心，北临珠江，南面有较大规模的储备用地。它将为人们提供一个地标性的集散、休闲、娱乐的滨水空间。

整体景观设计为自然式景观风格，运用弯曲的线条、简洁的材质表现自然现代景观的简约和生态。空间布局强调景观的参与性，尽可能创造多处临水活动空间。植物配置延续了琶洲河边绿化带景观，与周边生态环境和谐统一。公园设计还演绎了琶洲水乡文化，为周边的商业氛围注入更多历史人文景观，提升价值。

waterfront platform 滨水节点平台

landscape lake 景观湖

ecological streams 生态溪涧

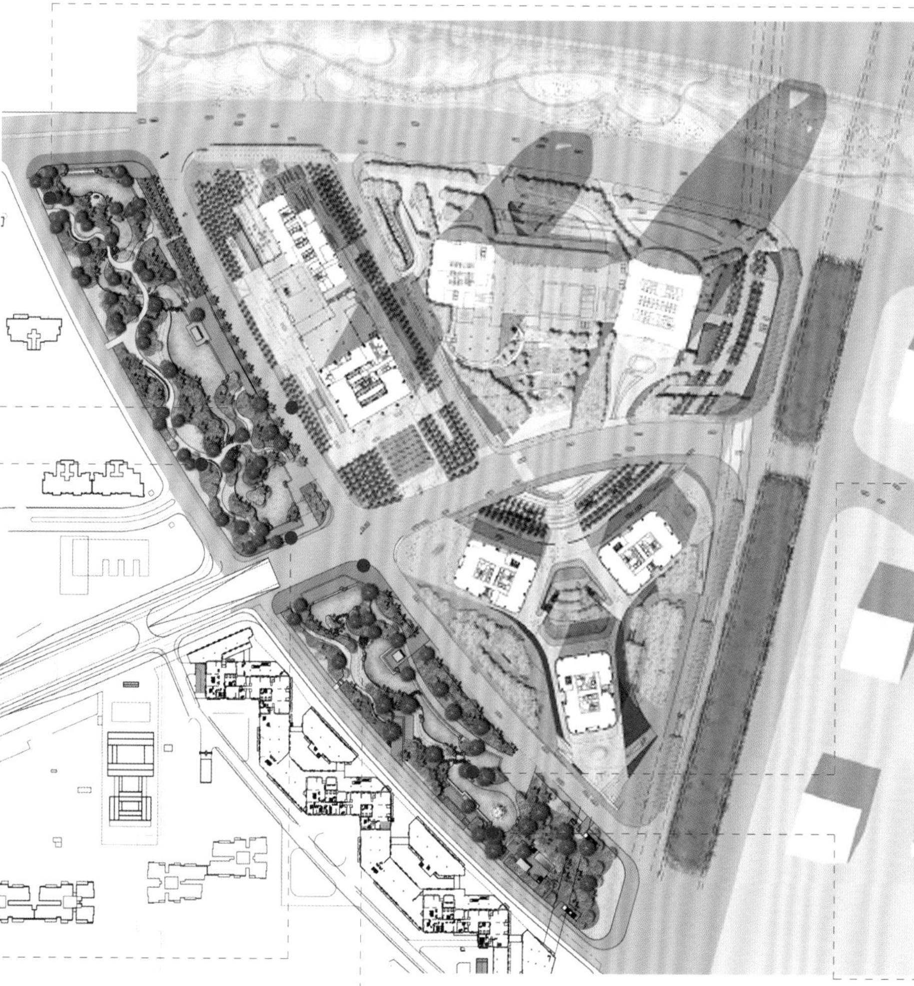

waterfront platform at the northwest corner
西北角滨水节点平台

ecological streams 生态溪涧

waterfront spaces 亲水空间

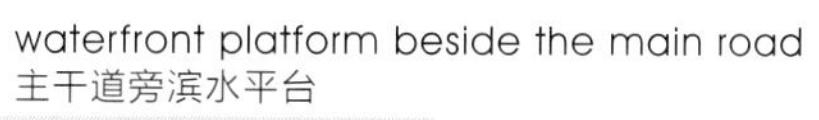

waterfront platform beside the main road
主干道旁滨水平台

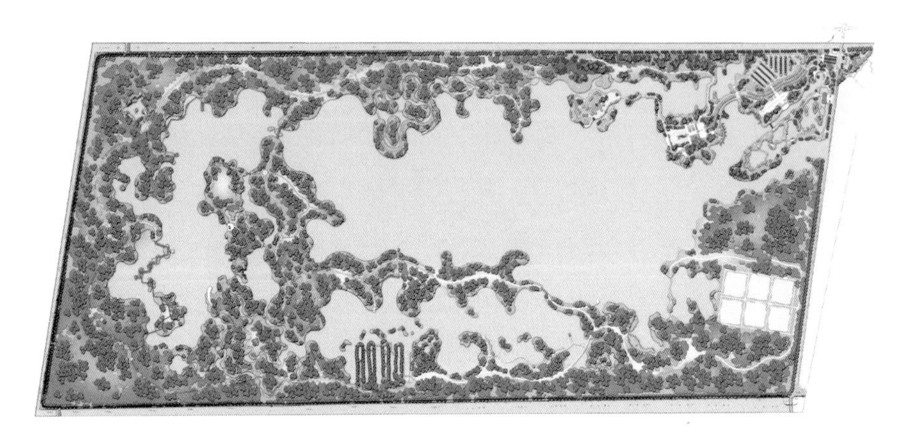

1
2 | 3

1 site plan of the entrance square and wetland experience zone
2 bird's-eye view of water bay square
3 bird's-eye view of wharf

1 南沙湿地入口广场及湿地体验区平面
2 水湾广场鸟瞰图
3 湿地二期游船码头鸟瞰

Nansha Wetland Park, Guangzhou
南沙湿地公园，广州
——围垦之生态湿地，仅存之鱼鸟天堂

Developer: Guangzhou Nansha Weiken Development Co.,Ltd.
Project Type: Urban Development
Project Area: 207,070 m^2
Design Content: Landscape Conceptual Design & Production Design
Design Period: 2012

委托单位：广州南沙围垦开发有限公司
项目类型：城市开放空间
项目面积：207 070 m^2
设计内容：景观方案及施工图设计
设计时间：2012 年

Nansha Wetland Park is located at the south end of Guangzhou, on the west bank of estuary of Pearl River. With the biggest wetland area in Guangzhou and the ideal ecological environment for birds' perch and reproduction, it has become one of the most famous tourist destination in Guangzhou and China. Adhering to phase one's design theme of "human, bird, sea and sky", the design for the entrance squares (phase one and two) and the landscapes (phase two) follows the idea of "being ecological, environment-friendly, natural and cultural" to create a diversified wetland area for education, scientific research and tourism.

南沙湿地公园位于广州最南端，地处珠江出海口，是广州市面积最大的湿地，也是候鸟的重要栖息场所，已经成为广州乃至全国著名的景区之一。根据发展需要，一、二期入口广场和二期湿地景区景观设计将顺应一期湿地“人、鸟、海、天”的总体理念，以生态、环保、自然、注重人文景观的设计核心，围绕“生态湿地、岭南水乡风情”构筑整个景观骨架，打造一个集科普教育、科学研究、旅游观光为一体的多样化的湿地景观旅游圣地。

1 rendering of the wetland plank road
2 rendering of the lotus area
3 eco, natural, beautiful and comfortable landscape in the experience zone

1 湿地栈道效果图
2 旭日观荷景区效果图
3 体验区景观生态自然、优美舒适

Nansha Wetland Park, Guangzhou

Panlong Mountain Park, Queshan
盘龙山公园，确山
——栖之山林，隐于世外

Developer: Shenzhen Urban Planning & Design Institute Co.,Ltd.
Project Type: Urban Development
Project Area: 550,000 m^2
Design Content: Landscape Conceptual Design & Production Design
Design Period: 2014

委托单位：深圳市城市规划设计研究院有限公司
项目类型：城市开放空间
项目面积：550 000 m^2
设计内容：景观方案及施工图设计
设计时间：2014 年

Located in the south of Queshan County, Henan Province, Panlong Mountain is near to Huang-Huai Plain on the east and leans against Mont Tongbai and Funiu on the west, boasting winding paths, jagged rocks, deep ravines, fragrant flowers, singing birds, as well as lights and shadows. Surrounded by the east lake, west lake, south lake and dragon lake, it appears like a piece of colorful satin embroidered with green mountain, blue lakes, gurgling streams and bubbling springs.

Enjoying advantaged landscape resources, Panlong Mountain Park is planned to be a comprehensive park integrating waterfront recreation, activity celebration, science education, cultural activities, sports and health, etc. Based on the natural topography, the landscape design has integrated with local culture and activities to create a 4A-class tourism resort in Panlong Mountain. It has restored the eco environment, diversified the local species, reorganized the site and recreated the waterscapes to realize the sustainable development of the landscape. Upon completion, the park will be an ideal place for recreation, relaxation and sports.

盘龙山位于河南确山县城南部，东临黄淮大平原，西依桐柏，伏牛两山余脉，其山势逶迤，岚影沉浮，峰石俏丽，林壑清幽，鸟语花香，霞光掩映，东湖、西湖、南湖、龙湖环抱其中，溪流潺潺，泉水叮咚，宛若一块刺绣着青山绿水的彩色绸缎，自然风光十分诱人。

该项目为老城规划集滨水休闲、活动庆典、科普教育、文化活动、体育健身为一体的综合性公园。整体以山水地形为特色，融入地方文化与活动，未来将对接整个盘龙山，打造 4A 级旅游度假区。此次改造主要是修复现有生态环境，丰富乡土物种的多样性，整理和设计场地地形，重塑水体，促进场地景观弹性发展，包括场地的景观形态和可持续发展。整体规划公园功能，满足市民休闲、游憩、运动等功能，挖掘场地特色，打造独具吸引力的活力体验公园。

1
2

1 bird's-eye view of the waterfront square
2 waterfront square

1 滨水广场鸟瞰图
2 滨水广场人视图

1 | 3
2 | 4

1 bird's-eye view of the entrance square
2 entrance square
3 bird's-eye view of the rainwater garden
4 rendering of the amphitheatre

1 入口广场鸟瞰图
2 入口广场人视图
3 雨水花园鸟瞰图
4 阶梯剧场效果图

Panlong Mountain Park, Queshan

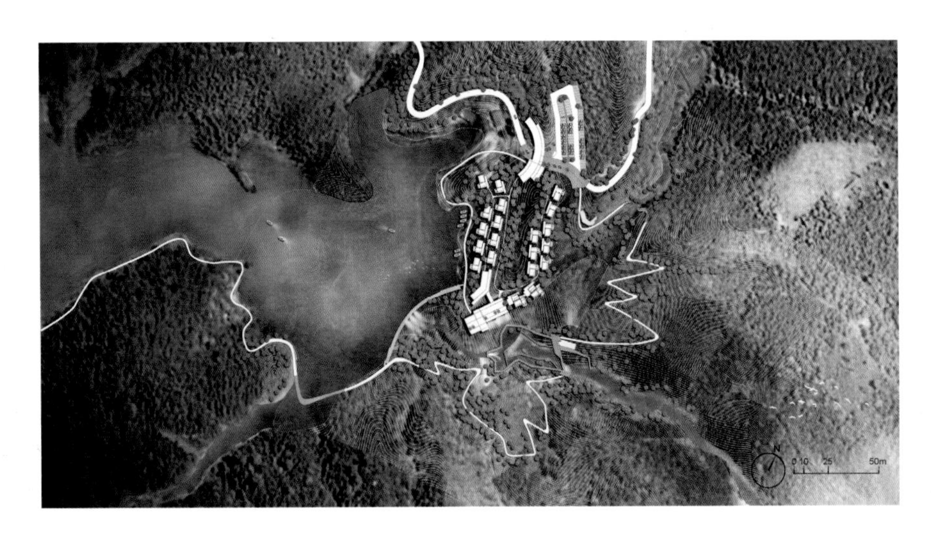

Daxiong Mountain Gaofeng Lake, Xinhua
大熊山高峰湖，新化
——临悬崖观绝世瀑布，登高峰揽峡谷胜景

Developer: Xinhua County Tourism Investment Development Co., Ltd.
Project Type: Urban Development
Project Area: 240,000,000 m^2
Design Content: Landscape Conceptual Design & Production Design
Design Period: 2015

委托单位：新化县旅游投资开发有限公司
项目类型：城市开放空间
项目面积：240 000 000 m^2
设计内容：景观方案及施工图设计
设计时间：2015 年

Gaofeng Lake Waterfall is located within the famous national forest park —— Daxiong Mountain Resort which is full of cliffs, waterfalls, streams, valleys and forest.

Taking advantage of the natural landscape and resources of Daxiong Mountain, it tries to present a modern, stylish and high quality destination in Central Hunan. Modern materials such as rust plate, light steel and glass are employed to create a series of thrilling footpaths and platforms, allowing tourists to enjoy the great cliff and waterfall from different perspectives.

高峰湖瀑布位于湖南新化县赫赫有名的国家级森林公园——大熊山景区内。景区内场地地形丰富，有悬崖、瀑布、溪涧、深谷、森林等，景观优质资源集中，景观基质极好。

从整个大熊山现有的同类型旅游资源出发，着力将项目打造成为整个湘中地区现代、前卫特色、极具传播效应的高品质旅游特色景点。景观设计上运用极具现代特色的锈板、轻钢、玻璃等材质打造出一系列刺激性极强的游步道节点和平台，让游人能够多角度、多方位地体验悬崖瀑布，在森林里遇见一场瀑布盛宴。

bird's-eye view　鸟瞰图

1	2	4
	3	5

1 observation platform
2 boulevard
3 waterfall
4 bird's-eye view of the waterfall
5 waterfall view from the observation platform

1 特色观景挑台
2 林下步道
3 远眺瀑布
4 鸟瞰飞瀑
5 观景平台观瀑

Daxiong Mountain Gaofeng Lake, Xinhua

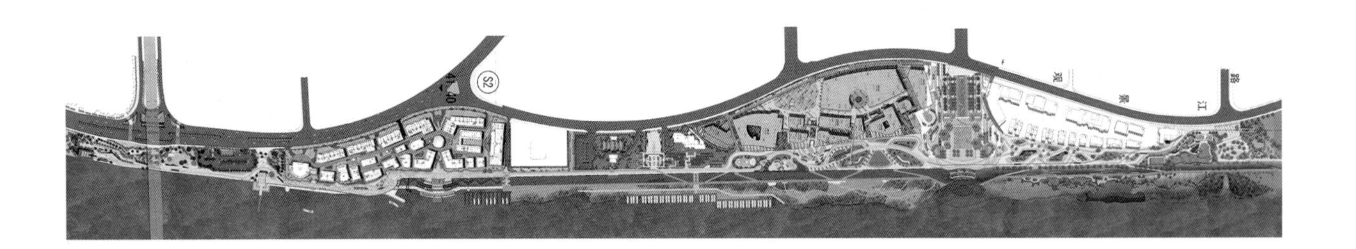

West Riverside Commercial Tourism Landscape Zone, Changsha

滨江新城西岸商业旅游景观带，长沙

——最聚国际风尚的都市滨江旅游目的地

Developer: Changsha Costal Newtown Development LLC
Project Type: Urban Development
Project Area: 143,178 m^2
Design Content: Landscape Conceptual Design & Production Design
Design Period: 2014

委托单位：长沙市滨江新城建设开发有限责任公司
项目类型：城市开放空间
项目面积：143 178 m^2
设计内容：景观方案及施工图设计
设计时间：2014 年

Sitting on the west bank of Xiangjiang River, Riverside New Town is the core demonstration area in building a "resource-conserving & environmental-friendly society" in Changsha. With ideal location, rich natural resources and the longest (8km) riverside green land, it is going to be the new landmark of Changsha city in the future.

Our design is for the 228km long and 10 ~ 85m wide landscape zone which extends from Tanshan Road to Yinpenling Bridge, on the east of the riverside landscape road. Lying along Xiangjiang River, it dialogues with "two pavilions and one hall" across the river. The commercial & tourism landscape zone on the west bank will become an international riverside tourist destination which integrates commerce, tourism, recreation, culture and sports.

Core Ideas: Dynamic and Ecological

Taking advantage of the abundant resources and local culture, based on the core ideas of "dynamic and ecological", it will create more attractive products and service as well as complete supporting facilities for this tourist destination.

Five Characteristics and Clear Theme

1) landmark: By exploring the local cultural characteristics and embodying them in function design, landscape planning and landscape architecture design, it aims to create an unforgettable destination in the site. Riverside green belt connects with urban roads to be the window to showcase the townscape of the new town.

2) participation: varied public spaces and waterfront activity spaces are created for people to stay and relax themselves.

3) accessibility: considering the accessibility to the surroundings, a multi-level and comprehensive traffic system is established to connect the riverside landscape zone with surrounding urban spaces.

4) diversity: according to the overall planning, it creates different functional

spaces and organizes diversified activities on different festivals to attract people. And the design of vegetation pays attention to seasonal changes and ensures different sceneries throughout the year.

5) ecological: it largely preserves the existing plants and uses environmental-friendly materials such as water permeable bricks to create an eco and sustainable environment.

1 | 2
3

1 beach and wooden platform
2 riverfront sightseeing tower
3 riverfront green land

1 休闲沙滩与木平台近景
2 临江观光塔，提供多重观景视点
3 滨江绿地设计潇洒而自由

滨江新城西岸商业旅游景观带，长沙

滨江新城位于长沙市区湘江西侧，是长沙建设“两型社会”的核心示范区，场地内 8 公里滨江景观带，是长沙现存沿江最长的可开发性滨水绿地，得天独厚的地理位置与自然山水，使之成为长沙未来城市新地标。

本项目位于最核心的 228 公里，东西宽 10 ~ 85 米，北起坦山路，南至银盆岭大桥，滨江景观道东侧，毗邻湘江，与“两馆一厅”隔江相望。西岸商业旅游景观带将成为最具国际风尚的都市滨江旅游地，一个集商业、旅游、休闲、文化、运动为一体的复合型商业旅游产品。

“活力”、“生态”两大核心理念：

整合湘江西岸商业旅游带的优良资源以及丰富多彩的人文底蕴，对资源进行深度挖掘，丰富旅游产品内容，完善旅游开发配套设施，围绕“活力”“生态”两大核心理念，构建丰富的产品层次，打造旅游带的核心吸引力体系。

1 | 2 | 3 | 4

1 overlooking the commercial tourism area on the west bank
2 multi-level observation deck of post modern style
3 square and observation deck
4 sail-shaped lamps along the river

1 临空俯览西岸商业旅游景观带
2 复层观景平台极具后工业现代感
3 集散广场与观景平台聚焦视点和人气
4 江岸沿线风帆造型的特色灯具

五重特性，主题突出

1）标志性：充分挖掘片区内资源的地域文化特色，并在功能分区、项目策划和标志性建筑景观上予以充分体现。打造标识性景观形成场地的记忆点，将城市通道与滨江绿化带进行有效的引导与连接，使之成为展示新城风貌的窗口。

2）参与性：在场地中增加不同的公共空间，打造六种形式的亲水岸线，植入多样化的活动功能，吸引人在此停留并进行活动，激活场地活力。

3）通达性：考虑滨江绿化带与外部交通的接驳，内部场地多流线交通的连续性，岸线上下层的立体连接，以及绿道将整个滨江新城空间网络化，打造多层次复合型交通系统。

4）多样化：根据规划营造不同的功能空间，在滨江绿化带组织出各种各样吸引人气的活动，植物配置方面丰富多彩，强调季相变化，并随着不同的节日举办不同主题的市民活动，建立活力可持续的旅游、公共活动目的地。

5）生态化：在现状基础上进行改造提升，保留可利用的苗木资源，采用透水砖等环保材料，建立湿地形成良好的可持续循环生态系统等，打造两型社会示范区。

GVL SELECTED WORKS
怡境景观作品集

更多资讯，请关注：

公司官网 www.gvlhk.com

新浪微博：“@GVL怡境景观”

微信公众号“GVL怡境景观”

欢迎扫描“GVL怡境景观”微信公众平台二维码

向辛勤付出的GVL全体成员致敬！

香港 广州 深圳 长沙 北京 上海

图书在版编目（CIP）数据

GVL 怡境景观 /GVL 国际怡境景观设计有限公司，佳
图文化编 . — 天津 : 天津大学出版社，2016.9
ISBN 978-7-5618-5160-9

Ⅰ . ① G · · · Ⅱ . ① G · · · ②佳 · · · Ⅲ . ①景观设计 Ⅳ .
② TU986.2

中国版本图书馆 CIP 数据核字（2014）第 194996 号

责任编辑 油俊伟

出版发行 天津大学出版社
地　　址 天津市卫津路 92 号天津大学内（邮编：300072）
电　　话 发行部：022—27403647　邮购部：022—27402742
网　　址 publish.tju.edu.cn
印　　刷 广州市中天彩色印刷有限公司
经　　销 全国各地新华书店
开　　本 250mm × 250mm
印　　张 30.5
字　　数 386 千
版　　次 2016 年 9 月第 1 版
印　　次 2016 年 9 月第 1 次
定　　价 388.00 元